Quantizing emotions

(The basic mathematics of psychology)

First Edition

Quantizing emotions

(The basic mathematics of psychology)

First Edition February 2021

Includes bibliography and references

mohammadhasanalgarhy@gmail.com

Contents:

Chapter 03: The proposed quantum mechanical approach & applications

Chapter 04: Evolution of emotions

Chapter 05: Mathematical psychology & psychophysics

Chapter 06: Epigenetics as a good candidate

Chapter 07: New concepts introduced & future research.

Introduction:

Science is the crawling of humanity towards the truth.

This book is presenting two main approaches to quantize emotions. It starts with basic and intuitive thought experiments to build up the logic used to propose these approaches. It then mentions some of the earlier techniques to measure and quantify emotions, including a brief about mathematical psychology and psychophysics.

The first approach is based on the hormonal-neural response to emotions and proposes a value for it, "The HN-value." The second approach is through the science of epigenetics.

Generally, a background in physics or mathematics is not required. You can still enjoy the book as the thought experiments are built in an analogous and intuitive approach that leads to the book's concepts. A physics or mathematics background would be a plus in few sections of the book.

The attempts proposed in this book to quantize emotions are built on previous scientific work and concepts from different science fields. The book presents few thought experiments to build up and explain the reasons behind proposing these specific methodologies and approaches to quantizing emotions. It is important to note that the book does not claim to be proposing a scientific theory. It is rather opening doors for further research and scientific validations.

Thought experiments and analogies had been the most successful human approach that led to the most significant discoveries in humankind's history. Throughout history, the greatest minds used thought experiments and analogies as a tool to understand the laws of the universe work, drive theories, laws, and equations.

Some of the most famous thought experiments are the ones that led to discovering and driving the laws of gravity, relativity, and quantum mechanics. The key has always been asking the right questions. It is not necessarily always the correct answer. But the right question is the key to opening more doors in science, philosophy, and any field of study.

Isaac Newton's famous question led to gravity "Why the apple fell on the ground while the moon did not?". The question seems to be quite basic and intuitive, yet a simple right question was asked, and a new era in physics started.

Einstein had always utilized thought experiments to understand and explain his revolutionary physical concepts as the famous train and lightning experiment for explaining special relativity. The accelerating elevator and person falling off the roof were some of the thought experiments used to explain general relativity.

Schrödinger's cat is another thought experiment used to criticize the Copenhagen interpretation of quantum mechanics and the state of quantum superposition. Yet, it is often used in theoretical discussions related to quantum mechanics interpretations.

The thought experiments, methodologies, and even the whole approach to quantizing emotions presented in this book do not claim to be confirming or presenting specific scientific theories; the primary intention is to open doors for potential scientific validation and discussion.

This book is presenting as well what could be the potential future research in various areas of science based on the proposed approach and methodology. This is not claiming to be giving the right answers but rather asking some questions that could be useful and inspire some new ideas and research.

Even though "The basic mathematics of psychology" is in the book's title, it does not claim to be re-inventing the wheel in that field. There is already a quite sophisticated field of "Mathematical psychology and psychophysics," which there will be a brief about in this book. This book aims to build on previous works from various science fields, presenting a new approach and opening the door for further research and scientific validation.

Can we consider psychology & quantizing emotions a science?

This question has been an ongoing debate almost since psychology started. The argument of not considering psychology as a science is mainly based on the idea that psychology is often missing experimental conditions and conclusive results; after all, how can you measure emotions or quantize them. How can you experimentally test the study of the human mind, qualities, and emotions?

A widely acknowledged definition of science is systematically observing a phenomenon or group of phenomena, then using those observations to develop laws and principles through developing hypotheses and experimenting with them.

Relying on statistics to test a hypothesis is an acknowledged science methodology; a famous example is the Higgs Boson. Particle physicists cannot directly observe the Higgs Boson; instead, they rely on statistics to know it is there.

Psychology is the study of the human mind and behavior. Psychology research could demonstrate that it follows the scientific method. A widely acknowledged approach for the scientific method is:

1. Observation and description of a phenomenon or group of phenomena.
2. Forming a hypothesis to explain the phenomena.
3. Make a prediction based on the hypothesis.
4. Performance of experimental tests of the predictions by several independent experiments.
5. Iterate: use the results to make new hypotheses or predictions.

Psychology research can start by observing patterns in human behavior. Psychologists usually develop a specific, testable hypothesis about why that behavior happens, or they might create a scenario and see if it leads to a particular behavior. Suppose they want to find the correlation between variables that a specific event and behavior are related to; in that case, they conduct a field experiment where psychologists carefully subject in normal, uncontrolled circumstances. If they want to determine causation –whether or not a particular event triggers a behavior – they will create a rigorous, highly controlled, and replicable laboratory experiment. Then they will use statistics to analyze relationships within those data to make sure the findings are reliable; the experiments are often repeated under the same conditions.

Still, how can you objectively quantify or even agree on a definition of something like happiness? It can be interpreted in too many ways; it is too abstract to be accurately tested. People can define happiness differently depending on their culture, circumstances, or even the momentary conditions they are facing. So how can researchers objectively define it?

Another point is that how can you reliably reproduce the results in psychology? People, their behavior, and even their perception of emotions vary from one place to another, from one era to another. Even the same person's behavior and perception can change with age, location, and conditions.

What psychologists mainly attempt to do is to operationalize abstract concepts like happiness or anger. They create, validate, and test a functional definition that serves as a good substitute for something abstract. For example, they might study happiness by tracking how often a person smiles or laughs. Or they can have people rate their happiness from 0 to 10. Another technique is to track the number of endorphins in the body or the neural response.

An important note about psychology is that it is not looking to capture a universal human experience because that does not exist. As mentioned earlier, people's behavior and even perception of emotions vary from one place to another, from one culture to another, from one era to another, and even the behavior and perception of the same person can change with age, location, and conditions.

Creating general strict rules for human behavior would miss many nuances and details in different people and situations. Variables, including differences between people, cultures, eras, and ages, should be acknowledged in psychology research. What could cause skepticism is that it is difficult to separate fact from the vast amount of pseudoscience, self-help books, advice columns, and websites making false claims about psychology. Managing to follow a systematic, objective methodology with hard evidence and testable hypothesis would support psychology as a science with beneficial, measurable outcomes.

Psychologists usually say everything psychological is biological. Which is something we are going to see over and over again in this book.

This book will follow the scientific method and how psychologists apply it to their research. As mentioned above, how can we agree on a definition for such abstract concepts? - In this book, I propose an approach that the physiological effects of emotions, for example, the hormonal and neural responses to emotions, could help us quantize emotions by measuring them. I propose a process to measure and quantize them through few thought experiments and proposed mathematical approaches.

Some psychologists think that a set of ideas borrowed from quantum mechanics could help us make sense of human behavior. The notion is called quantum cognition, suggesting that quantum mechanics' mathematical tools could help make human behavior more predictable. This book will follow some of the quantum cognition methodologies, which will help us in our approach to quantizing emotions.

Chapter 01

Thought experiments

Introduction to the basic mathematics of psychology

The correct answer to one plus one

Several pages were written by the well-known mathematicians and philosophers Bertrand Russell and Alfred North Whitehead in their book Principia Mathematica attempting to prove the proposition 1+1=2.

This thought experiment suggests that to prove that 1+1=2 is not possible philosophically.

The reason is simple 1+1 is not always necessarily equals to 2 in nature. Suppose a man married a woman and got one kid. That is 1+1=3 if they got two kids that are 1+1=4; if they get five kids, that is 1+1=7.

Suppose we claim that we cannot follow this logic to describe 1+1. In that case, we have an issue with our current mathematics, which is the inability to express or, in other words, quantify certain situations. Otherwise, how would I quantitatively describe this situation? As at least philosophically, a man marrying a woman could be considered as 1+1.

Is mathematics reality, or is it describing reality? A big question that has been approached in many different ways. This example simply shows that mathematics does not always describe reality.

We have two main alternatives to handle it; the first one is to develop a new system to describe some aspects of reality or work on our mathematics to be adaptive enough to acknowledge different ways of describing reality.

Another example will be if we are planting a seed into the soil. We could consider the seed as one, the soil as one, and the result – a tree, for example- as one. We can follow a different logic: to consider the seed as one, the soil as a matrix of components leading to an ecosystem –the tree- which we could view as a matrix of new components. We could view the seed as an operator acting on the soil matrix, and the tree is the resulting matrix or ecosystem or even potentially few matrices. The logic of operators acting on matrices is already followed in mathematics and physics.

Technically, the seed and the soil are composed of atoms and elementary "particles," which are basically waves or collapsed wave functions. On a side note, quantum mechanics is evolving -and it should- beyond the wave-particle duality. However, it is acknowledged now that "quantum particles" are actually not exactly particles or objects; they are always waves or collapsed wave functions or quantum excitations of a field or waves of these fields tied up into little bundles of energy. In other words, both the seed and the soil are fundamentally waves. For further clarifications on this point, please refer to the "What is a particle?" section in this book.

So if the seed and the soil are waves, the logic can continue that an operator of a wave function is acting on a matrix of another wave function(s).

However, our standard addition, subtraction, division, and multiplication are not enough to describe, track and acknowledge the full operation. We should use a different approach. Let's move on to the next thought experiment to see if there could be another way. We do have Schrodinger's equation, which helps us understand how the wave function evolves in time. We will start by following an intuitive primitive logic to eventually reach the concept of the approach discussed in this book.

The "IN" sign

We have developed mathematics and physics where humans are basically out of them. We know that you will get a yellow color when mixing a red color & a green color. But can we measure the effect of the result on the perceiver? We have some guesses on the psychological impact, but we do not have a tool to measure it. As explained in the previous thought experiment, 1+1 is not necessarily always equals to 2 in nature, yet our current mathematics fails to describe these cases quantitatively.

In this thought experiment, let us consider a new sign called the "IN" sign (please bear with me, it is just a thought experiment attempting to explain the approach). Then, let's check if the "IN sign" can help explain and quantize events and situations in nature that could not be quantified by our current mathematical and physical operations. And I will use the notation ✖ for the "IN" sign. It is just a notation to help explain the idea.

Let's do few examples and thought experiments to examine it.

Example 01: Let's start with a very primitive and intuitive illustration of the seed and tree from the previous thought experiment "The correct answer to one plus one." Let's consider X as the seed and S as the soil, and the result T refers to the tree. So, a seed in soil equals a tree.

$$X ✖ S = T$$

However, this is almost meaningless. It fails to define many factors and qualities for the seed, the soil, and the tree. It does not tell basic information like at least what type of seed it is. For example, if we know it is a lemon seed, at least we have one piece of information that says something. At least we can predict that if it is a lemon seed in the soil, we expect the probability of getting a lemon tree to be 100%. Yet this is very basic and intuitive. How could this help?

9

Let's consider that we require more specific information; we need to know more details about the resulting tree and fruits' qualities. We need to know the tree's height, the fruit's size, the tree's productivity, and the tree's life span. We can still guess this from our previous experience of the same type of seed and soil in a similar environment. However, we cannot quantize it. Let's consider a matrix of qualities of the seed and a matrix of soil qualities.

$$\begin{vmatrix} x_1 & x_2 \\ x_3 & x_4 \end{vmatrix} * \begin{vmatrix} s_1 & s_2 \\ s_3 & s_4 \end{vmatrix} = \begin{vmatrix} T_1 & T_2 \\ T_3 & T_4 \end{vmatrix}$$

Let's consider X_1 as the type of the seed; X_2 is the size of the seed X_3 & X_4 as two other qualities. S_1 as the soil type, S_2 as the fertilizers used in the soil, S_3 & S_4 as two other qualities. The addition or multiplication of these two matrices will not serve the purpose or give us the information we are looking for. It is required to find a way where all the qualities of matrix one can operate on all the qualities of matrix 2.

But what would the "IN product" be? It has to be predefined by the one carrying the operation. So, for example, we should predefine that T_1 will result from the operation of X_1 on S_1 & S_2 and so forth. This approach will help track the process, quantize it, isolate, and introduce factors and qualities.

Let's move on to **Example 02**; we will use the analogy from the previous thought experiment, "The correct answer to one plus one," about a man getting married to a woman. Here the qualities become much more, and the factors that we can consider are huge. We can even track it down to the DNA level to analyze this example's "IN product." The resulting matrix will be probabilistic, but the predefined conditions will be required to define it. With some computational measurements of probabilities, we can predict almost all the operation's potential outcomes with a high accuracy level.

$$\begin{vmatrix} GS_{1W} & GS_{2W} \\ GS_{3W} & GS_{4W} \end{vmatrix} * \begin{vmatrix} GS_{1M} & GS_{2M} \\ GS_{3M} & GS_{4M} \end{vmatrix} = \begin{vmatrix} GS_{1K} & GS_{2K} \\ GS_{3K} & GS_{4K} \end{vmatrix}$$

Let's consider a fundamental operation where a woman's genome sequence is in a man's genome sequence to get the kid's resulting genome sequence. In this example, we considered four different genome sequences of women in four different genome sequences of a man. There could be thousands or even millions of possible results of this operation. The addition or multiplication of these two matrices will not serve the purpose. We should find a way where all the qualities of matrix one can operate on all the qualities of matrix two and the other way around as well where the qualities of matrix two operate on the qualities of matrix one.

Again, this is just a hypothetical thought experiment to explain the approach; however, other genome sequencing and DNA testing methods exist.

We will take this to the next step by the following thought experiment, "Quantizing emotions."

Quantizing emotions – A thought experiment

This thought experiment is based on the previous two thought experiments, "The correct answer of one plus one" and "The IN sign," which are considered a prerequisite for this thought experiment.

The big question is, how can you quantize qualities? As mentioned in our thought experiment "The IN product," we have developed mathematics and physics where humans are basically out of them. We know that you will get a yellow color when mixing a red color & a green color. But can we measure the effect of the result on the perceiver?

Following the logic in the previous two experiments, let's consider **Example 03** for two men supporting two football teams; the first one is supporting the team with a red uniform. The second one supports a team with a blue uniform. Let's start with the simple basic equation.

$$R \divideontimes M_1 = EM_1$$

This equation tells us that the color red "IN" the first man will give us emotion #01; we can define this further that generally, when this man sees the red team, emotion 01 would be happy. We can call this the **first state**. We have a very high probability that this equation is correct, and the first state would be happy.

We can add a condition here, whether the team won or lost. In this case, the results could be different, but it will be relativistic, which means if the first man, the red team supporter, saw his team losing, the emotion or state could change to sadness or depression. However, this could be relativistic as well since one supporter can get extreme sadness, anger, and depression, and another supporter may perceive it in a more chilled way. The reaction could depend on the match's importance as well, either it was in the finals of an important championship or just a friendly game.

But with all these factors and qualities, how can we quantize them?

I would like to introduce here a few concepts and definitions:

1. The unique factor.
2. Pre-defined conditions.
3. Imposed conditions.
4. Complex states.
5. Isolation principle.

Before we jump into definitions and explanations, I have to clarify that the IN sign ✳ can accept any operator, such as colors, shapes, sounds ....etc. Let's stick to the color in our example and see how we can quantize it. There are few ways to quantize it, and I will mention here two direct methods to approach it: the hormonal response and the neural response; both can be measured.

Studies show that we can measure changes in certain hormone levels in the body as a response to social challenges. This response has been interpreted as an adaptive mechanism that helps individuals adjust their behavior to social context changes. Studies show an increase in the testosterone levels in the winning teams' fans and decreased testosterone levels in the losing teams` fans.

In an attempt to a bit visualize it:

Color, shape, sound ✳ ▶ Hormones / Neurotransmitters ▶
Measure & quantize

Back to the definitions and explanations of the introduced concepts:

We will call the first part of the equation (red in this example) the operator, the second part of the equation (man in this example) the operatee, and the result is the "IN product." In some cases, the operatee can become the operator and vice versa. They can switch roles.

1. **The unique factor** is a relativistic factor/constant that differs from one operatee to another.

   Explanation: The unique factor measures the average hormonal-neural response for a specific period. In our example, the unique factor would be his usual or average level of reaction to the win or loss of his team. We can quantify this reaction by the relevant hormone or neurotransmitters levels; for example, we can take an average value due to these reactions, and a unique numerical factor could be assigned. This unique factor can change from one age to another for the same operatee. The unique factor can change from one operation to another based on the information we want to drive. More examples will follow in the next two thought experiments, "The quantum party" & the "The quantum ride."

2. **Pre-defined conditions:** A group of conditions specific to each operation should be pre-defined to provide the necessary information to carry out the operation and predefine the required information to be shown in the "IN product."

   Explanation: while the unique factor defines criteria about the operatee, the pre-defined conditions define criteria about the operation and the expected outcome. In our example, let's consider pre-defined conditions of an important game in the finals of a championship. The operatee is a big fan of the red team, and we want to measure the increase or decrease of the testosterone levels in his body after the game.

3. **Imposed conditions:** A condition or more that is unconventional or unexpected, when imposed on the operation, will lead to a change in the results (the IN product)

   Explanation: In our example, let's consider an accident that happened to the red team fan. This condition is again relativistic; if it is a small accident, the fan can still go to the stadium to watch the game, and it will not affect his mood or the rise of testosterone in his body. If it is a big accident, the operation results (the IN product) could change. Imposed conditions could lead to canceling the operation.

4. **Complex states:** It is a number of factors/states that we need to consider that could affect the hormonal-neural response and lead to different "IN product."

   Explanation: In our example, we want to measure the red team's effect of scoring the first goal on a young red fan between the ages of 18-25. We want to measure the increase in the testosterone levels in his body, neural response, and the effect on his productivity levels, for instance.

   Let's try to write this down:

   $$R_t \, G_1 \, \divideontimes \, Y_r M_r = TN$$

   R_t is the red team, G_1 is scoring the first goal, Y_r is the young condition of 18-25 years old, and M_r refers to the first operatee, the fan of the red team. The "IN product" would be the H level of changes of testosterone levels in his body and N, the neural response of his body. This could be written as matrices similar to those proposed earlier in the "The in sign" thought experiment.

The ⚹ sign gives any operator the freedom to act on any operatee based on the pre-defined conditions and the required outcome of the operation (the IN product).

Let's perform the same operation on a blue team fan:

$$R_t\,G_1 \divideontimes Y_r M_b = TN$$

Rt is the red team, G1 is scoring the first goal, Yr is the young condition of 18-25 years old, and Mb refers to the operatee, the blue team fan. The "IN product." would be the H level of changes of testosterone levels in his body and N the neural response of his body. We expect a very different hormonal-neural response from the red team fan. Potentially we will get opposite results.

We can run the operation for a specific game; in this case, we will be able to measure the hormonal & neural responses directly and compare them before and after the game.
We can run the operation on different games and conditions to get the estimated average hormonal and neural response.

5. **The isolation principle:** It is the principle of isolating specific states or conditions from the operation to get the required pre-defined results of the "IN product."

   Explanation: The isolated factors should not be affecting the core operation or the "IN product.". For example, we cannot isolate the condition of winning. Still, we can isolate the color of the uniform whether it's home or away as we predefine that it will not have that much effect on the operatee.

The quantum party

This thought experiment is based on the previous three thought experiments, "The correct answer of one plus one," "The IN sign," and "Quantizing emotions," they are considered a prerequisite for this thought experiment.

Let's consider a group of four young friends going to a party. Our pre-defined conditions are the age group that they are all going there willingly and happily. The "IN product" is required to measure the dopamine, serotonin, endorphin, and oxytocin levels in their bodies individually and collectively.

Let's consider that as state 01 with the pre-defined conditions mentioned above

A simple form would be: $P \divideontimes G = HN$

P stands for the party and H for the hormone or neurotransmitters` levels, G stands for the group of friends, but this will not satisfy our pre-defined conditions. We need to be more detailed as follows:

$$P \divideontimes \begin{vmatrix} F_1 & F_2 \\ F_3 & F_4 \end{vmatrix} = \begin{vmatrix} HN_1 & HN_2 \\ HN_3 & HN_4 \end{vmatrix}$$

F_1 to F_4 refers to the four friends; we isolate each one's age as it will not affect the predefined conditions. P is for the party as we do not need to detail it more. HN_1, HN_2, HN_3 & H_4 are for dopamine, serotonin, endorphin, and oxytocin, respectively.

As per our pre-defined conditions, we were required to measure these hormone or neurotransmitter levels for each friend individually and then collectively. In an attempt to do that:

$$P \divideontimes \begin{vmatrix} F_1 & F_2 \\ F_3 & F_4 \end{vmatrix} =$$

$$\begin{vmatrix} HN_{11} & HN_{21} \\ HN_{31} & HN_{41} \end{vmatrix}, \begin{vmatrix} HN_{12} & HN_{22} \\ HN_{32} & HN_{42} \end{vmatrix}, \begin{vmatrix} HN_{13} & HN_{23} \\ HN_{33} & HN_{43} \end{vmatrix}, \begin{vmatrix} HN_{14} & HN_{24} \\ HN_{34} & HN_{44} \end{vmatrix}$$

Where HN_{11} is the dopamine level for the first friend, HN_{21} is the serotonin level for the first friend and so on.

We can also measure the average hormonal and neural response on all the operatees.

The analogy goes that our bodies are formed of atoms, and as mentioned in the thought experiment, "The correct answer to 1+1", we are fundamentally waves. Given that, I am considering in this thought experiment that the four friends have separate wave functions and a collective wave function tuned based on their state during and after the party. We are attempting to measure the quantum state of this wave function individually and collectively. Again this is just a thought experiment attempting to explain the approach proposed in this book.

Let's impose a condition, one of these friends came to the party angry and intending to ruin the party. That friend can ruin the party and lead to a totally different "IN product," or he can decrease happiness and, accordingly, the hormone levels. How do you know the effect? We measure it following the logic introduced above.

Let's try to write it down:

$$P \divideontimes \begin{vmatrix} HF_1 & HF_2 \\ HF_3 & AF_4 \end{vmatrix} =$$

$$\begin{vmatrix} HN_{11} & HN_{21} \\ HN_{31} & HN_{41} \end{vmatrix}, \begin{vmatrix} HN_{12} & HN_{22} \\ HN_{32} & HN_{42} \end{vmatrix}, \begin{vmatrix} HN_{13} & HN_{23} \\ HN_{33} & HN_{43} \end{vmatrix}, \begin{vmatrix} HN_{14} & HN_{24} \\ HN_{34} & HN_{44} \end{vmatrix}$$

HF_1, HF_2 & HF_3 refer to the happy friends & AF_4 refers to the angry friend. Following the same logic introduced above, we can measure each friend's hormone levels individually, and then we can average the measurements to measure them collectively. Accordingly, we can estimate that angry friend's effect on the party.

The quantum ride.

This thought experiment is based on the previous four thought experiments, "The correct answer of one plus one," "The IN sign," "Quantizing emotions, "and the "Quantum party," they are considered a prerequisite for this thought experiment.

We will consider a woman's daily ride to her work, listening to her favorite music, and measure her hormonal and neural response: M ⋇ W_v = HN.

M is for the music, W_v is for the woman in her vehicle, and the "IN product" would be the hormonal and neural response (HN).

Suppose these predefined conditions remain unchanged and without imposed conditions for a month. In that case, we can measure her hormonal and neural response daily, and we can average it across the month as well for pre-defined standard conditions.

Let's impose some conditions; for example, one day she is in a rush, the other day she had an issue with her kids, the third day her radio was not working. In other words, there could be several reasons for mood disorders during this ride. Applying the isolation principle introduced earlier, we can isolate and measure every reason/operator's effect, causing mood disorder and disturbing the "IN product" of this daily ride. Once we can isolate these operators' effects and quantize them, we will be able to figure out solutions to fix them if they negatively impact and enhance them if they have a positive impact quantitatively.

Following the logic introduced in the previous four thought experiments, our woman is a quantum wave basically listening to sound waves. We are attempting to measure the effect of the operator (the sound waves) on our operatee (the quantum woman).

We have pre-defined conditions that this music makes this woman happy; let's impose new conditions where the woman changes her work or location &, for some reason, stopped listening to this music. Then let's consider that an unfortunate situation happened to her. For example, breaking her leg, let's consider an operation of $BL \divideontimes W_v = HN_2$. We have a new operation here, with a new numerical outcome of the hormonal and neural responses. The question is: can we use the numerical result of the $M \divideontimes W_v = HN_1$ to help her mitigate the negative effect of breaking her leg? It is not that simple, though.

We will have to consider the concepts proposed earlier:

1. The unique factor.
2. Pre-defined conditions.
3. Imposed conditions.
4. Complex states.
5. Isolation principle.

For example, the woman's unique factor would be the average level of her response generally to unfortunate events. We would measure the response here by the hormonal and neural effects. This measurement would vary from one person to another. One person can be having extreme depression, anger, etc. Another person can take it more lightly and deal with it. We could average out the unique factor on a span of three years, for example. Again, this is pretty relative and could vary from one person to another. So our predefined conditions would tell us whether we need to consider the unique factor for six months, for example, or three years.

Let's give the unique factor a scale from 0 to 10. It should have a sign in front of it, so for example, -2 indicates that unfortunate events have extreme adverse hormonal and neural effects on that person, and +2 means that this person has a way even to turn unfortunate circumstances into a positive mindset. Again, this could be averaged from the hormonal and neural response through 3-5 years, for example, or as we pre-define.

Consider a numerical value for the HN (Hormonal and neural response). This value would be the product of the level of the hormones or neurotransmitters we want to track and the unique factor. For example, if we found out the hormones or neurotransmitters levels over one month average out to 50 ng/mL and that the woman's unique factor is, for example, 1.5, then the HN-value would be 75 HN. We can also measure the HN value per day and get 75 HN/day. Further details on calculating the hormonal-neural response and the unique factor will be explained in chapter two, which is "The proposed mathematics." Generally speaking, we can check the effect of UF (Unique factor) on the average hormonal and neural response of the first operation.

UF ⵣ HN_1

We can have a unique factor in perceiving good events and a separate unique factor for sensing unfortunate events if required to apply isolation principles. We can also average them out to have a general unique factor.

Suppose the woman broke her leg, and we made her listen to the music she loves or imagines the happy ride, or even go for the same ride listening to her favorite music. Would that mitigate the negative effect of the unfortunate event?

Even if the answer might seem obvious, it is pretty probabilistic, yet it could be experimentally done. Imposed conditions might have a saying in that, for example, how severe was her leg broken, were there other unfortunate events happening at the same time, what did breaking her leg prevents her from doing. Probably, if the leg was not severely broken and no other imposed condition, her happy ride would refresh her and improve her HN effect, while if a loved one died during the same period, not only probably the happy ride will not have that much positive effect. Actually, the new operation of these unfortunate even might ruin the memory of this happy ride forever if we tried to use it. In other words, an event/operation in the present can affect an event/operation in the past.

Could that be measured? The answer is yes, and the above-proposed model can help to understand this and the probabilities of these effects.

For example, if we measure the HN_2 response of the second operation $BL \maltese W_v = HN_2$ and we tried to ask the woman to go for the happy ride and measure the new hormonal and neural response after the effect of the second operation. We can tell the impact of the second operation on the first operation if we keep carrying on the same experiment several times over the years after becoming well and overcoming her loved one's loss. Suppose we measured the average hormonal and neural responses and compared the results before the unfortunate events. In that case, we will tell whether the woman completely "psychologically" recovered or the operation of the regrettable events is still affecting her. Yet, we will not lose hope; if we have a record of the woman's hormonal and neural responses, we can try to predict and sort of design the best solution to overcome the situation in the long term and even improve her hormonal and neural response.

Chapter 02

The proposed Mathematics

The HN-value (Hormonal –neural value)

In the thought experiments presented in chapter one, we noticed the emergence of the HN-value (Hormonal –neural value), which is a method to quantize emotions based on the hormonal and neural response to emotions. It is based on the two-system view of the stress response, which is explained in further detail later in this book and explains why this method, "The HN-value," is quantitatively more reliable than other methods and attempts to "measure emotions." A unit could be **HN/day,** for example.

But how would the numbers work?

There are three main methods:

1. The most straightforward method would be the direct measurement of the levels of selected, pre-defined hormones and neurotransmitters responding to emotions and then getting the product of these levels and the unique factor. This is the main method we will follow and explain in this book.
2. Another method could be assigning scales for the hormonal response, the neural response, the unique factor, etc. If the hormonal-neural response falls within a specific range, then we assign values on a 0 to 10 scale, for example, and so on.
3. Another approach -which could eventually yield more specific and fundamental results- is studying how these responses evolve with time, similar to how Schrodinger's equation tells us how a wave function of energy evolves in time. However, this needs a lot of work to be developed and more than one mind to develop it.

We can use the first method (the direct measurement of the levels of selected, pre-defined hormones and neurotransmitters) in simple basic operations. Consider, for example, the sports game mentioned earlier; we want to measure the dopamine and testosterone levels for 2 participants from the winning team and two participants from the losing team; the calculation would be straightforward.

As mentioned above, the second method of assigning scales for the hormonal response, for the neural response, etc., could be useful in specific situations, especially if we want to get more abstract results and particular indicators.

The third method is considered one of the main targets of this book, is to open the doors for further research in this regard. The analogy could initially follow the methodology of quantum cognition and consider a model comparable to Schrodinger's equation, telling us how a wave function of energy evolves in time. We target to know how the effect of emotions evolves over time. The analogy could go that emotions are the energy operator, and we are the wave function. Further explanation about this approach in chapter three, "The proposed quantum mechanical approach & applications."

The two-system view of the stress response

When the body is exposed to excessive harm or threat, this initiates physiological changes known as the stress response. These are adaptive changes in response to the stressor that is beneficial in the short term but eventually have adverse effects on an organism. This response is centered around the anterior pituitary-adrenal cortex system, better known as the HPA axis. Stressors stimulate the release of adrenocorticotropic hormone from the anterior pituitary, which triggers glucocorticoids' release from the adrenal cortex. These are the signaling molecules that produce many physiological effects on the stress response. The sympathetic nervous system is involved in releasing epinephrine and norepinephrine from the adrenal medulla.

Combining these two systems is called the two-system view of the stress response. Stress responses are complex, and they vary from person to person. Still, we know that specific common psychological stressors relating to jobs, relationships, or other social obligations can have effects similar to physical stressors.

Psychoneuroimmunology (PNI) is a field of study that looks at the interactions between your central nervous system (CNS) and your immune system. Researchers know that our CNS and immune system can communicate with each other, but they only recently started to understand how they do it and what it means for our health. The nerves in your brain and spinal cord make up your CNS, while your immune system is made up of organs and cells that defend your body against infection. Both systems produce small molecules and proteins that can act as messengers between them. In your CNS, these messengers include hormones and neurotransmitters. On the other hand, your immune system uses proteins called cytokines to communicate with your CNS.

The two-system view of the stress response as a basis for the HN-value (Hormonal-Neural code) proposed in this book.

The two-system view of the stress response acts as a very good basis for the HN-value or the Hormonal-Neural code proposed in this book to help quantize emotions. The hormonal and neural response mechanism explained gives a reliable quantitative basis which avoids the disadvantages of other methods like the PANAS, mDES, SAM, ESM, FACS, SPAFF ...etc

It does not involve participants "describing" or rating their emotions, which is not entirely reliable quantitatively for obvious reasons. Some of them are explained in the quantum cognition section of this book. Neither does it involve analysis for facial expressions, which is not a comprehensive approach as well. Rather it gives us an accurate and direct physiological measurement of the effect of emotions on our bodies.

We can derive the HN-value's numerical value from the hormones and neurotransmitters' test reports. The key is to be consistent and use the same method and values with the same participant.

For the Central Nervous System (CNS), we can look at responses using functional magnetic resonance imaging (FMRI). Here we can look at the brain areas that seem to be activated during different kinds of emotional challenges or stressors.

The idea here is that in response to emotions, both hormones and neurotransmitters are released, which is why in this book, studying and measuring both the hormonal & neural response and assigning a value for it is proposed to help measure and quantify emotions.

So what is the Hormonal-neural response?

Neurotransmitters are chemical messengers that transmit a signal from a neuron across the synapse to a target cell, which can be a different neuron, myocyte, or gland cell. Neurotransmitters are chemical substances made by the neuron specifically to transmit a message.

We have got over a hundred different kinds of neurotransmitters. Some examples of neurotransmitters are endorphins linked to pain control and pleasure. Norepinephrine helps control alertness and arousal; it is an example of excitatory neurotransmitters. Glutamate is involved in memory. Serotonin that affects your mood, hunger, and sleep is an example of inhibitory neurotransmitters. Low serotonin levels are linked to depression, and a specific class of antidepressants helps raise the amount of serotonin in the brain. Some psychologists say everything psychological is biological, and neurotransmitters are good proof.

Some neurotransmitters can excite or inhibit neurons depending on the type of receptors they encounter. For example, acetylcholine which enables muscle action and influence learning and memory. Alzheimer's patients experience a deterioration of their acetylcholine-producing neurons. Dopamine is another example associated with learning, movements, and pleasurable emotions. An excessive amount of dopamine is linked to schizophrenia and addictive and impulsive behavior.

Hormones are chemical messengers manufactured by the endocrine glands, travel through the bloodstream, and affect other tissues. Hormones affect our mood, arousal, circadian rhythm; they regulate our metabolism, monitor our immune system, signal growth, and help with sexual reproduction.

The nervous and the endocrine systems are similar in that they both produce chemicals destined to hit up specific receptors. However, they operate at a very different speed. The endocrine system is much slower than the nervous system for specific reasons and functions.

Some of the hormones are chemically identical to certain neurotransmitters. And the key point here is to measure the levels of these chemicals, whether neurotransmitters or hormones, to quantify emotions.

In the case of stress, stressors stimulate adrenocorticotropic hormone release from the anterior pituitary, which triggers glucocorticoids release from the adrenal cortex. These signaling molecules produce many physiological effects on the stress response. The sympathetic nervous system is involved in releasing epinephrine and norepinephrine from the adrenal medulla.

As mentioned earlier, the idea here is that in response to emotions, both hormones and neurotransmitters are released, which is the reason why in this book, studying and measuring both the hormonal & neural response and assigning a value for it is proposed to help in measuring and quantizing emotions.

How do we usually "measure emotions," and what fits into the model proposed in this book?

One of the approaches is measuring the subjective experience of the person feeling the emotion. "How do they feel," then we'll look at how people measure the outwards expressive display of emotional behavior; in other words, how do they look when they feel emotional. Finally, we will conclude by discussing physiological ways of responding, so physiological ways to measure emotions in the autonomic and central nervous system, answering the question of what the body is saying when I am feeling a given emotion.

Measuring the subjective experience of emotion: When we think about subjective measures, we can roughly break them into three broad categories: Self-report questionnaires, continuous rating dials, and experienced sampling methods to capture emotion in everyday life.

One of the common questionnaires is the Positive & Negative Affect Schedule (PANAS). Where there is a host of different emotion-related adjectives that participants are asked to rate on a specific scale, they can answer this in terms of what they feel right now, what they felt in general in the past week or at a trait level, how they generally feel.

Another tool is called Modified Differential Emotional Scale (mDES). It comes in triplets of emotions, and people are asked to indicate to what extent from not at all to extremely, for example, hopeful/optimistic/encouraged to, for instance, love/closeness/ trust.

Another commonly used questionnaire measure is called the Stress Appraisal Measure (SAM). It uses graphical depictions of related emotional intensity compared to using emotionally related adjectives, which could have a different interpretation from one person to another.

Another method is called narrative methods, capturing more nuanced themes of a person's unique emotional life. One of the things proposed in this book is quantizing the narrative method throughout a person's lifetime.

The narrative method collects people's verbal or written descriptions of their emotional experiences, and you can code it along a rich dimension of different categories. For example, you can look at the event's temporal sequencing, sort of what's the linear trajectory or lack thereof in which they tell the story. You can look at the particular kinds of emotional words or metaphors people use. You can also look at the rich context they embed their emotions in and whether they describe their emotions as active agents over or simply a passive recipient. These are things you cannot get using questionnaires. However, you can get using a narrative methodology.

We move from questionnaires and narrative methods to what is called rating dials; they allow you to look at fine-grained second-by-second measurements of emotion. So what you have people do is while they're in a laboratory, you have them place their hand on a dial where they can move it from very negative to very positive, for example. This has been used in studies where people are either watching dynamic movies over several-minute periods or interacting in discussions with partners. People can shift the dial second by second as their emotional experience changes.

The third sort of common measure of subjective experience is the "Experience Sampling Method (ESM)," a method to collect information about a person's emotional state outside the laboratory in their own naturalistic daily life. An example of that is mobile apps that people can repeatedly report on a given day at random intervals. How are they feeling, what context are they in, what kind of strategies are they using to effectively manage their emotions?

Now let's move to behavioral methods, how do we look, how do we behave when feeling emotional. Two of the commonly used tools are the Facial Action Coding System (FACS) and the Specific Affect Coding System (SPAFF).

The Facial Action Coding System (FACS) was one of the first systems developed to quantify emotional expressions in the face systematically. It was developed by Ekman & Friesen in the 1970s. It tried to categorize facial behaviors not based on overall sort of subjective impressions of how emotional the person's face looked but instead based on a more micro-level rigorous coding of the underlying muscle configurations that are used to give rise to an emotional expression. So every little muscle action unit is what he called an action unit or AU. He would argue that specific AUs combine to reflect emotions. The numbers in codes like 1C, 5D, 4B, etc., are the action units, and the letters next to them refer to the intensity of the display. So they begin from A, which is sort of the lowest end, all the way up to E, which is the maximal of how much you could contract that muscle. For example, AU6 "The cheek raiser" & AU12 "Lip corner puller" is used to identify a real smile.

The Specific Affect Coding System (SPAFF) includes additional codes that aren't inherent in the FACS system, developed to help measure emotional behavior in the context of dyadic interactions.

We have talked about subjective and behavioral. What about our bodies? How does our body say anything about how we feel emotionally, and how do we measure that? Some of the methods to measure emotions in the lab where you have a variety of electrodes or sensors. They pick up the heart's electrical activity every time the heartbeats and look at different respiration patterns, which are then graphed; you can map specific types of Autonomic Nervous System (ANS) changes to emotional responses. So we can try to map specific physiological profiles of anger, fear, sadness, disgust, and even happiness based on changes in our bodily responses. So that is the peripheral autonomic nervous system.

For the Central Nervous System (CNS), we can look at responses using functional magnetic resonance imaging (FMRI). Here we can look at areas of the brain that seem to be activated during different kinds of emotional challenges or stressors.

What hormones or neurotransmitters are we looking for, and how can we measure them?

Since hormone and neurotransmitters levels are the main tools proposed to quantize emotions in the "HN value approach," we need to understand what hormones and neurotransmitters are and what types of hormone and neurotransmitters tests are out there. Which hormones or neurotransmitters and which test results are we looking for to measure the effects of emotions and events on our hormone levels and carry out the mathematical operations mentioned in this book.

Hormones are chemical messengers that are released mainly by glands, and they travel in the bloodstream through all parts of the body to targeted cells. Hormones then tell these cells what to do. Water-soluble hormones bind to receptors on the surface of cells, the message is delivered, and the cell does its duty. Fat-soluble hormones enter and bind to receptors once inside, the message is delivered, and the cell does its duty.

For example, glucagon is a hormone produced by the pancreas. After we eat, carbohydrates are broken down into glucose; the liver sends some of this glucose into the bloodstream to cells throughout the body. The cells use this glucose as energy. The liver takes any extra glucose and stores it as glycogen or as fat. As this glucose in the bloodstream gets used up by cells, the liver needs to release more, and this is where glucagon comes in. The pancreas releases glucagon hormones, which attach to receptors on liver cells. Glucagon tells the liver cells we need more glucose, and the liver cells convert the stored glycogen into glucose and release it into the bloodstream.

There are many different hormones in the body that have many different functions. There are hormones that help with blood pressure regulation, the development of white blood cells in response to stress, regulation of water retention, and the kidneys. Hormones serve to communicate between organs and tissues for physiological regulation and behavioral activities such as digestion, metabolism, respiration, tissue function, sensory perception, sleep, lactation, stress induction, excretion, growth and development, movement, reproduction, and mood manipulation. Hormones have diverse chemical structures, mainly of three classes: Eicosanoids, Steroids, and proteins /amino acids.

Neurotransmitters are chemical messengers that transmit a signal from a neuron across the synapse to a target cell, which can be a different neuron, myocyte, or gland cell. Neurotransmitters are chemical substances made by the neuron specifically to transmit a message.

As psychologists say, everything psychological is biological.

Hormones and neurotransmitters are associated with the regulation of psychological aspects like stress inductions and mood manipulation. To quantize emotions, we need to consider all sorts of physiological regulations and behavioral activities, including things like sleep, growth and development, metabolism, etc.

We need to apply the isolation principle introduced in this book to determine which physiological regulation or behavioral activity we want to deal with. We need to predefine conditions before every operation we carry to predetermine which hormones and neurotransmitters we are targeting to measure.

Here we are not discussing a case of low melatonin hormone levels (a high level of melatonin allows you to sleep, while a low level keeps you awake). Physicians prescribe some synthetic melatonin for you to assist you in sleeping. Instead, we want to mathematically track the underlying events, circumstances, and events that led to low melatonin or any other hormones causing depression or anxiety disorders and mood disorders in general. Then develop the mathematics to deal with it through the principles introduced in this book. For example, imposing & isolating conditions that we now have a very high probability of dealing with the situation, potentially avoiding deteriorated psychological and even medical conditions.

The next section of this book will overview the hormones and neurotransmitters tests, the test results, and the numbers we could use for this book's purpose.

What numbers or test results should we consider, and how is that integrated into the proposed equations?

Let's have a general overview of how hormonal tests work. There are few types of hormone level tests. For example, the saliva test is a non-invasive method where the patients collect their saliva in plastic tubes to measure hormones like cortisol, estrogens, progesterone, and androgens. Another testing method is serum tests performed by drawing blood from a patient and spinning it down to separate the liquid serum's cellular components, containing soluble analytes such as hormones. 24-hour urine collection is another testing method that is usually preferred for testing hormones secreted at night and during deep sleep, such as growth hormone 14 & melatonin 15.

While saliva, serum & 24-hour urine tests are the most popular tests for measuring hormones, few other techniques and companies compete to innovate testing methods with higher accuracy and reliability. And of course, every testing method has its pros and cons and is usually preferred for specific conditions and requirements.

Some test methods could generally prove to be more accurate than others. What is important is the consistency of using the same testing methods, the same numerical set of values, and units with the same client. It is like whether you decide to use miles or kilometers to measure distance. However, a testing method could prove to be more convenient and reliable for our purpose.

The hormone test results are usually shown in mIU/ml (milli-international units per milliliter), which is pretty familiar in HCG-pregnancy hormone test results. Sometimes the hormone levels are given per number of hours, for example, mIU/ml/6h, which is pretty convenient. Other units that could be seen in some test results are ng/ml (Nanogram per milliliter) or µg/L (micrograms per liter), or pmol/L (Picomoles per liter). In some situations as well, the units could be micrograms per liter (μg/l), micrograms per 24 hours, μg/24h) and micrograms per kilogram per 24 hours (μg/kg/24h). For dosage, mcg/kg/minute is often used.

So, here let's go for the gram per liter (g/L) unit. It is a mass per volume kind of operation. The same applies to neurotransmitters; the neurotransmitters' levels are usually tested through urine tests, and you per mass per volume results like ng/mL or pg/mL.

Let's take a simple example from our quantum ride thought experiment, and let's measure the dopamine levels in the operatee.

$$M \maltese W_v = HN$$

M is the music operating on W, which is the woman in her daily ride resulting in an HN-value (Hormonal-Neural level value), which for this operation would be just the dopamine level.

We will measure the dopamine level daily; let's consider we found it in the range of 0 to 30 pg/ml (Picogram per milliliter) (195.8 pmol/l picomoles per liter).

Now let's consider that this woman is a fairly happy person in general and has a unique factor (UF) of 1.5; let's apply this to the equation introduced earlier.

$$UF \maltese HN_1$$

In our case here, it will be just the product so, let us consider on a day her dopamine level was 20 pg/ml with a unique factor of 1.5. This will lead to an HN-value of 30 HN. Let's impose an even happier condition of a call she receives that day telling her that she got a promotion. We test her dopamine levels and find it 40 pg/ml, 60 HN on that specific day. Let's impose some sad news for her and measure the dopamine level to find it two pg/ml, which is 1.5 HN. Let's consider a complex state of receiving a promotion and hearing some bad news about a beloved one and measure again and so forth. This is a rather very basic situation, but to explain the approach in simple terms.

What if we want to measure more than one hormone and neurotransmitter?

In that case, we will take the value of the hormones and neurotransmitters and divide them by their number. For example, if we have two hormones, we will divide by two, or we have two hormones and one neurotransmitter divided by three. But to be consistent, we will have to show how many hormones and neurotransmitters were considered in the operation. And to do that we could show it in the unit so for example.

$$M \ast W_v = HN \ (2H1N)$$

For example, if we have two hormones with 100 pg/ml and one neurotransmitter with a value of 50 pg/ml. The total would be 150/3 = 50 pg/ml by the unique factor of 1.5. This will give us 75 HN (2H1N). The predefined conditions will tell us which two hormones and one neurotransmitter we are considering in this operation.

If we applied that to a group of friends as in our quantum party thought experiment and measure the dopamine levels for each of them, it will look like that.

$$P \ast \begin{vmatrix} F_1 & F_2 \\ F_3 & F_4 \end{vmatrix} = \begin{vmatrix} HN_1 & HN_2 \\ HN_3 & HN_4 \end{vmatrix} = \begin{vmatrix} 35 & 30 \\ 40 & 25 \end{vmatrix} \text{pg/ml}$$

And then we can move on with:

- Isolating & imposing conditions.
- Applying complex states and unique factors.
- We can average the results of the full group to get specific outcomes.
- We can measure more than one hormone & neurotransmitters, then integrate them into the equation.
- We can predefine the conditions and required outcomes.

For example, an average value for this operation would be (35+30+40+25)/4 = 32.5 pg/ml, which will use an average unique factor of 1.2, which will give us 39 HN.

We can also **normalize the unique factor** by considering it one.

As simple as it initially looks, the equation gives us a tool to carry out different operations types that could help us better understand, quantize and sort of control the situation or at least propose recoveries or enhancements. The matrices give us a great tool to add operators and operatees. The IN allows swapping roles, as explained in this book.

It will not be that simple, though, or might not even yield integer results when we apply the "IN sign" to Schrodinger's equation—coming up in the next chapter.

The power of the HN-value

To demonstrate how the HN-value could potentially be very powerful to track and sort of control emotions mathematically

Let's compare three women with an HN-value of 65 HN (1H2N) & 15 HN (1H2N) & -10 HN (1H2N). What do these numbers and units tell you?

Well, if you have just opened the book on this page and you saw these three numbers, most likely your impression would be ok there is a high number, a smaller number, a negative number, then more numbers and letters.

This unit says that we have measured 1H (One hormone) & 2N (two neurotransmitters) & multiplied these numbers by a unique factor for each woman who indicated how she generally responds to specific emotions.

Let's give some more information to measure Melatonin, serotonin & dopamine. We wanted to understand specific factors related to sleeping behaviors & disorders related to depression, joy, and mood.

Then you might start having an impression that the first woman's unique factor (which refers to how she generally responds to events or emotions) is higher than the other two women. It seems that the third woman has pretty adverse or sort of pessimistic reactions in perceiving situations and events.

And you are right; the HN-value's power is that it can give you some excellent indicators about the person and his reactions to emotion. Based on that, we might need to conclude that either the unique factor or the level of hormones or neurotransmitters need to be improved for the second woman and dramatically improved for the third woman.

How can we change the unique factor and level of hormones or neurotransmitters?

Physicians and psychologists follow many methods, but our goal here is to measure and track these methods and their results mathematically. Let's say we want to improve the unique factor (how they generally respond to events or emotions) of the second and third women. And we decided to start training them to perceive & respond to events and situations differently and more positively; we follow a training program and measure the HN value daily (HN/day). We average the value on six months, for example, then measure the unique factor and see if there is an improvement in the unique factor and the levels of hormones and neurotransmitters.

We need to design the action we will take, predefine conditions, isolate conditions, impose conditions and create a complex state if required. We need to predefine the durations required to measure the unique factor and the HN-value. We can design the needed action to create an impact over one month or six months, or even a year. Designing the action depends on the situation.

Measuring the unique factor

So how can we measure the unique factor?

The unique factor would vary from one person to another. We could average out the unique factor on three years, for example. We are considering that within 3-5 years, a person's hormonal and neural response could have significant changes.

We should predefine a set of unique factors depending on which outcomes we target. For example, we can define a unique factor indicating how the person is generally perceiving and responding to events and emotions, either being generally optimistic or pessimistic. We might then be looking for the hormones or neurotransmitters associated with depression, pleasure, etc. So we might be looking, for example, serotonin and dopamine.

So we measure serotonin and dopamine weekly over six months, for example, which will give us a pretty good idea of the unique factor.

We measured the average serotonin level weekly over six months. We found it 200 ng/mL. We measured the dopamine level weekly over six months. We found it 30 pg/mL, which is 0.03 ng/ mL; her HN-value would be (200+0.03)/2 = 100.015; to get the unique factor, we will divide by 100 to get 1.00015.

Let`s consider another case of average serotonin level of 110 ng/mL and average dopamine level 0.01 ng/ mL, the HN-value would be (110+0.01)/2 = 55.005 & the unique factor 0.55005.

We do not necessarily have to divide by 100 to get the unique factor; it depends on the reference ranges of the neurotransmitters or hormones which we have predefined; we need to get the unique factor in the range from 0 to 10, we might need to divide by 10, 100 or potentially 1000, etc.

We can normalize the unique factor by setting it to one. The unique factor plays an essential role in better understanding the operation and operatee. And the effects of the operator and the operation on the operatee.

Emotions measurement software

Software attempting to "measure emotions" is mostly based on using sensors. They synchronize sensors like eye-tracking glasses and face readers. Usually, it is measured in a laboratory; some companies attempt to measure it in the participants' daily lives. Measurements typically include heart rate activity, eye movements, facial expressions, etc., using various tools.

Some software and mobile apps have more advanced questionnaire tools. An example of these tools is participants can fill questionnaires about their emotions or report over multiple times in a given day at random intervals how they feel, what context they are in, and what strategies they are using to manage their emotions effectively.

There is a wide range of data outcomes, metrics, and analyses. They are tracking, for example, what sometimes is called an emotional pattern, emotional activation, etc. It could be categorized by age, gender, geographical location, etc. The data analysis and outcomes could be customized based on the requirements.

Yet, the vast majority of these software and apps are mainly relying on facial expressions, heartbeat, etc., which in concept is not very different from the methods mentioned earlier in this book. In other words, they are just digitalizing the already existing methods of "measuring emotions."

Chapter 03

The proposed quantum mechanical approach & applications

Introducing the "IN sign" to Schrodinger's equation

This part is a bit more technical; it would require some physics background. A sort of a disclaimer that I have to mention as a reminder, what I am proposing here is just an approach to the subject from a different perspective, and it will require scientific validation.

I will have to explain some technical background that led me to propose introducing the "IN sign" to Schrodinger's equation. The thought experiments presented in chapter one, "The correct answer of one plus one," "The IN sign," "Quantizing emotions, "Quantum party," and the "Quantum ride," are considered as a prerequisite for this part of the book.

Max Planck quantized energies in multiplies of $\hbar\omega$ and then came Einstein said the energy of a photon is $\hbar\omega$ and the momentum of the photon is $\hbar\vec{k}$. So $(E,\vec{p}) = \hbar(\omega,\vec{k})$. Then came De Broglie with the idea that even though this was written for photons, it was valid for particles as well, all particles and these particles are waves $\Psi(x,t) = e^{i(kx-wt)}$ Since k is positive, this is a wave moving to the right.

The operator that realizes the momentum would be
$$P = \frac{\hbar}{i}\frac{\partial}{\partial x}$$

The operator that realizes the energy would be
$$E = i\hbar\frac{\partial}{\partial t}$$

When the energy operator operates on the wave function

$$i\hbar\frac{\partial}{\partial t}\psi = \hbar w\psi$$

$$i\hbar\frac{\partial}{\partial t}\psi(x,t) = E\psi(x,t)$$

This equation is prescribing how a wave function of energy E evolves in time. It tells you if you know the wave function and it has energy E, the left side of the equation is how it looks later, and you can take the derivative on the left side of the equation and solve this differential equation. In this equation, E is a number. If you know you have a particle with energy E, that is how it evolves in time.

So came Schrodinger and looked at this equation, which is true for any particle that has energy E. Maybe I do not know what the energy E is. One single replacement was introduced to this equation, replacing the energy E with the energy operator $\hat{E}$. Schrodinger`s equation does not assume that the energy is a number because you do not know it. In general, if the particle is moving in a complicated potential, you do not know what the possible energies are.

$$i\hbar\frac{\partial}{\partial t}\psi(x,t) = \hat{E}\psi(x,t)$$

This is symbolically what must be happening because if this particle has a definite energy, then the energy operator $\hat{E}$ gives you the energy acting on the function.

Given that, what led me to introduce the "IN sign χ" to Schrodinger's equation is if the particle (basically wave or, in other words, collapsed wave function) that the energy operator is operating on does not have definite energy? What if the energy operator is operating on a complicated potential or, in other words, a field of waves, which is fundamentally how energy operators operate. They do not operate on a single wave. They operate on a field of waves.

The energy operator requires the freedom to act on the operatee's multiple quantum states. Both the operator and the operatee can affect each other, so this requires the freedom to swap roles as introduced in our thought experiment, "quantizing emotions."

Additionally, we would require to include other operations being acted upon by the energy operator, not just position and time. The other operatee could eventually probably be reduced to position and time. However, we will need to define more operatees during the operation.

$$i\hbar\frac{\partial}{\partial t}\psi(x,t) = \hat{E} \; \chi \; \psi(x,t)$$

This equation would be the simplest form of introducing the in sign to Schrodinger`s equation. The energy operator is acting "in" a wave function with definite energy, so basically, the IN sign here is doing nothing.

$$i\hbar\frac{\partial}{\partial t}\psi(x,t) = \hat{E} \divideontimes \hat{\psi}(x,t)$$

Here the equation is prescribing that the energy operator is acting "in" a wave function, which is the operatee in this case, and this wave function does not have definite energy; it could be a field of waves. This equation also prescribes that the energy operator and the wave operatee affect each other and can swap roles.

If we take the analogy from our earlier thought experiment, "The quantum party," And again, this would be a thought experiment. Let`s consider in this example the energy operators as the sound & color as energy operators of a complicated potential that we can drive. We want to isolate all the other factors and measure the energy operator's effect on dopamine levels, which is another complicated potential formed of atoms and subatomic particles/waves. The impact of the wave function of dopamine on the sound energy in the party location is negligible. Given that, it will not be required to swap operator and operatee roles. However, the operatee is still quite complex. It will require further isolation of factors, and in this case, defining it just in terms of position and time will not serve the purpose of the thought experiment or our required "IN product."

So far, we are trying to understand how the wave function evolves in time. What if we want to understand how the wave function evolves relative to other factors, shape, for example? This shape could be itself evolving with time and could be time-independent; these are pre-defined conditions that we should define before carrying out the operation. These pre-defined conditions will help us conclude whether we will require the first or second derivative of the "IN product" to get the information we are looking for, which will take us to our next thought experiment, "Energy in shape = A wave function."

Energy in shape = A wave function.

This could be considered as well as a thought experiment. It is based on the thought experiments presented in chapter one, "The correct answer of one plus one," "The IN sign," "Quantizing emotions, "Quantum party," "Quantum ride," Introducing the "IN sign" to Schrodinger's equation, they are considered as a prerequisite for this thought experiment.

In this thought experiment, we will examine the energy as an operator on or rather "in" shape, following the logic presented in the thought experiments introduced in chapter one. The energy will be the operator acting "in" shape, which is the operatee. However, they can swap roles. The "IN product" would be a wave function that could be defined in terms of position and time or any other parameters.

$$\hat{E} \divideontimes \hat{s} = f^{\Psi}(x,t)$$

This equation is how it potentially look like. Where $\hat{E}$ is the energy operator, $\hat{s}$ is the shape operatee and $f^{\Psi}(x,t)$ is the resulting wave function. This proposed equation prescribes that energy in shape gives a wave function, which could be applied in principle to almost everything. However, the five concepts mentioned in the thought experiment, "quantizing emotions," should be considered: the pre-defined conditions, the unique factor, the imposed conditions, the complex states, and the isolation principle in order to get more specific information and solve more specific problems.

The concept of the shape as an operatee or even an operator in this proposed equation goes beyond the conventional shapes to topological spaces construction, construction of new topologies from given topologies, topological manifolds, bundles, etc., is a big area that could be explored. The idea is that shape/space acts as a container for energy. Energy can operate on the shape and change it, and vice versa shape can operate on energy and change it is frequencies. The shape's concept goes even further to the idea of the pattern, which is discussed in this book.

The resulting wave function or "IN product" should be considered an operator. In this proposed equation, we have the freedom to swap the in "IN product," the operator, and the operatee. In other words, the energy operator could act on the wave function to give a specific shape or topological manifold, etc. in that case; the equation would be:

$$\hat{E} * f\Psi(x,t) = \hat{s}$$

Swapping the roles of the operator, operatee, and the "IN product" is a fundamental principle that could be useful in many conditions and various states.

This idea is not too exotic from the superstring theory. T- duality is interchanging W (the winding number) and n (the momentum). In my proposed equation $\hat{E} * \hat{s} = f\Psi(x,t)$, the energy operator $\hat{E}$ is comparable to momentum and the wave function is comparable to the winding number.

Can the shape of a drumhead be predicted from the sound that it makes, in other words, from the spectrum of vibrations? You can predict a lot of the shape of the drumhead. The same question applies to compactifications in string theory. From the spectrum of particles, in other words, the spectrum of vibrational energies, can you predict the shape and the size of the compact directions? The answer is yes to a large extent.

The question here goes beyond if we can predict the drumhead's shape from the sound it makes. The question is whether the spectrum of vibration can form the drumhead's shape? And whether both the shape and the vibrational energies can form certain functions or rather certain wave functions. How would that wave function evolve in time? And how would it affect and get affected by the surrounding ecosystem?

Here the five principles mentioned earlier in this book could be helpful:

1. The unique factor.
2. Pre-defined conditions.
3. Imposed conditions.
4. Complex states.
5. Isolation principle.

If we understand what could be referred to as the spectrum of vibrational energies of a specific situation, somehow manage it & "tune it" to get specific results, we would be achieving one of the main goals of this approach. But how can we sort of "decode" this spectrum of vibrational energies? Do we need frequency modulation?

Back to the idea that the shape in $\hat{E} \ast \hat{s} = f\Psi(x,t)$ could go further to the idea of the pattern, where the movement or rather the fluctuations of energy or complicated potential of energy in a particular pattern creates a specific function or a wave function, it could as well written as follows :

$$\hat{E} \ast P = f\Psi(x,t)$$

Where P refers to the pattern in which the energy operator moves; following the same logic, the operator can affect the operatee and vice versa.

Similarly, the freedom to swap the "IN product," the operator, and the operatee. In other words, the energy operator could act on the wave function to give a certain pattern; in that case, the equation would be:

$$\hat{E} \ast f\Psi(x,t) = P$$

An example of the geometrical pattern is what we can consider as the energy pattern in which the atoms are arranged in a molecule. This pattern is giving the molecule its function. The atom is formed of a specific pattern of electrons, and more fundamental "particles," the change in this pattern due to change in energy can change the function and lead to a different molecule. Let's go back to the idea that a particle could be considered a collapsed wave function. Could the collapse of the wave function target a specific pattern to stabilize and carry out the targeted function?

Back to $\hat{E} * \hat{P} = f\Psi(x,t)$. Even the energy operator in this equation might not affect & could be considered negligible in some situations. This means that only the pattern can lead to a specific function. This is analogous to the atom & molecule story. When atoms are joined in specific geometrical patterns, we get a particular substance's molecule. If the same types of atoms are joined in a different geometrical configuration or pattern, we get different kinds of molecules of other substances. In this case, the energy of the atoms played no role and the specific function/substance was formed just based on the geometrical pattern or configurations. In other words, the pattern led to a certain function, which could be expressed as $\hat{P} = f\Psi(x,t)$.

The effect of the patterns on the resulting molecule could be directly measured and understood as we can study which pattern of atoms leads to which molecules and hence which types of substance; accordingly, we can understand which patterns lead to which sort of "functions." Studying the effect of patterns is pretty much what we are attempting to do with emotions. The analogy goes that if we managed to understand which patterns of emotions lead to which functions or effects, this will get us one giant step closer to effectively quantizing emotions.

Every electron's location and energy in an atom are determined by a set of four quantum numbers that describe different atomic orbitals. An orbital is a region of probability where an electron can be found. These four quantum numbers are:

- "n" is the principal quantum number telling us the energy level.
- "ℓ" is the azimuthal quantum number that tells us the orbital type and angular momentum.
- "m_ℓ" is the magnetic quantum number, which tells us which specific orbital is amongst the set and the projection of the angular momentum.

- "m_s" is the spin quantum number that tells us the spin. Each electron in an atom has a unique set of quantum numbers. The quantum numbers describe the pattern and the effect of this pattern.

A small thought experiment considering two energy systems, for example, observer and observed energy system, are in resonance, exchange of information happens. Afterward, both will not return to their original state. They both keep part of the information, which causes change to both energy systems, and in a sense, they stay connected or entangled. This kind of entanglement is pretty much comparable to the act of measurement on a quantum system, where quantum entanglement happens with the measuring tool.

Should we consider emotions as energy operators or wave functions?

What could appear intuitive is that emotions could be considered energy, causing an effect or regarded as work. The analogy is that if humans are made out of particles and particles are collapsed wave function, or we will consider for this analogy as just wave functions or waves; then emotions are energy operators operating on a wave function evolving with time.

Yet, let me go back to this: $\hat{E} \ast \hat{s} = f\Psi(x,t)$

In that case, the analogy could be that emotions are an energy operator acting on or "in" a shape that we will consider as our human, for now, resulting in a function that evolves with time.

Yet, remember that the "IN sign" gives the freedom to swap roles for the operator, operatee, and even function. So the analogy could then be that emotions are wave functions, and our humans are energy operators that could affect this wave function. But how is that even possible?

Let's go back to our thought experiment, "The quantum ride.". Let's consider the sound waves of the music that makes our lady happy, as "waves of emotions." So far, they could be viewed as sort of energy operators acting on our operatee causing an effect or work which is the emotion of happiness. But we are not done yet; let`s consider one day that while she is listening happily to her music, she receives a call that one of her loved ones passed away. There is a significant probability that this music will not be the "waves of happiness" anymore; instead, it would be linked with a terrible memory. Next time and potentially every time she listens to this music, she will feel terrible. Even her HN (Hormonal- Neural response) to the very same music used to cause her happiness will probably change to probably a negative response.

This thought experiment tells us three significant conclusions (which are yet still analogous):

1. The first one is that humans could be considered as a measuring tool –basically, we are a very sophisticated measuring tool for the universe- that causes collapse of the wave function of emotions when measured. The emotions could be considered in a superposition, and once we measure them, they sort of taking of a specific position.

2. The same measuring tool -which is basically us humans in this case- could lead to different results due to various circumstances. It is not necessarily always a spin-up or a spin-down. It could be more complicated than that. We could consider that humans are a measuring tool that could be calibrated, and the calibration could be self-controlled as well.

3. Another analogy is that emotions could be considered either energy operators or wave functions, "waves of emotions."

The analogy will continue that in case there are no complex states or imposed conditions (please refer to the five principles mentioned earlier in this book) and the "energy of the sound waves of the music" is affecting our "Human collapsed wave function." Then emotions could be considered energy operators (it could be regarded as analogously as waves acting on waves). In this case, we would need to study the role of the measuring apparatus, which is basically us humans.

The second case is if we have complex states or imposed conditions that lead to changing the energy operator's effect. Basically and still analogously, the wave function of the measuring apparatus (us the humans) was sort of calibrated to perceive the "wave function of emotions" differently. The importance of the "IN sign" appears here and gives the ability to swap roles of these equations. Another perspective would be is that we as humans operated on the "wave functions" we are perceiving, and we change their effect from optimistic to unfavorable or less favorable. It could be considered that the operator became the operatee; in other words, they swapped roles.

This analogy brings us to another two remarkable concepts, "Patterns of emotions" and "The derivatives of emotions," which will be proposed in the next couple of sections in this book.

Patterns of emotions

Let's take this analogy to the next step, which is:

$$\hat{E} * P = f\Psi(x,t)$$

We now understand that this equation tells us that energy (in other words, energy operator) in a pattern gives a specific function. We know as well the concept of swapping roles. Following this line of thought, emotions could have a pattern, which could be referred to as "Patterns of emotions." However, it would not be patterns of shapes; it would rather be patterns of wave functions. Our target would be to find out the correct "Pattern of emotions" to cause the required positive effect on a wave function. Of course, it will not be a one-time effect. It will be something evolving with time. Which is something the above equation might help us figure out. Incorporating the "IN sign" in Schrodinger's equation could be an effective method as well.

But what is the pattern in this context? How can we quantify it?

One of the simple analogies in this context is patterns of music (sound waves), where specific patterns of these sound waves could cause an effect that evolves in time. Our target is to quantify that by measuring the operator's respective wavelengths and the HN value (Hormonal – neural response) on the operatee in a specific period. We could then have data and statistics to understand the situation better.

Music is just one example or analogy for the idea of how specific patterns can operate on a field of waves to produce a measurable effect. And how these patterns could be quantized and better understood. We would need to sort of decode various types of patterns. One technique could be translating every pattern into what we sense with our organs and could be broken down into wavelengths like sound waves and light waves/color.

Which brings us back to the question mentioned earlier in this book and presented in a lecture at Stanford University by the famous Physicist Leonard Susskind back in 2010: can the shape of a drumhead be predicted from the sound that it makes, in other words, from the spectrum of vibrations? You can predict a lot of the shape of the drumhead. The same question applies to compactifications in string theory. From the spectrum of particles, in other words, the spectrum of vibrational energies, can you predict the shape and the size of the compact directions? The answer is yes to a large extent, as mentioned by Leonard Susskind.

The derivatives of emotions

Let`s start with a quick overview of what derivatives are and why derivatives could be very effective when attempting to quantize emotions.

A derivative represents the rate of change of a function. A second derivative represents the rate of change of the function. For example, we can graph a position of an object versus time; this could be a falling object or a speeding car, for example. The derivative of this position function will be the rate of change in this function or the change in position with respect to time, and that is velocity. So velocity is the derivative of position. The derivative of this velocity function will be the change in velocity with respect to time, and that is acceleration. The position is in meters, and velocity is in meters per second, and acceleration is meters per second per second; in other words, meters per second squared (m/s^2). That means that acceleration is the second derivative of position.

The analogy is simple if we can consider the rate of change of one emotion as a derivative, and the rate of change of the rate of change is the second derivative. For example, we could consider that serenity as a derivative of happiness. Yet, we have to be quite careful here; to define which emotions are truly the derivatives of other emotions, we have to follow a reliable methodology. We need to make sure that every time there is a rate of change in this specific emotion, we get its derivative for sure. Statistics and surveys will not be the most effective and reliable tools for this purpose. Yet, the HN-value's actual measurements (Hormonal and neural response) would give us more reliable results.

Most researchers would agree on what is usually referred to as the basic or primary emotions to be six: Happiness, sadness, fear, anger, disgust, and surprise, which are differentially associated with three core effects reward (happiness), punishments (sadness), and stress (fear and anger).

To actually get reliable results on which emotions could be derived from the basic emotions or if we even need to change the classifications, this requires extensive research and experiments measuring the HN-value (Hormonal & Neural response). It could be considered with respect to time analogs to what we mentioned in position, velocity, and acceleration. So, for example, exposing a group of participants to a happy experience for a month measuring the HN-value per day, we could design the experience to measure the rate of change in the HN- value on a daily or weekly basis.

Repeating the experiments many times on different groups would give us an educated framework on which emotions could be derived from others, which emotions could be considered the second derivatives, and so forth. The HN- value would help us quantify that and assign actual numbers and units for them. A unit could be **HN/day,** for example.

However, an important note is that it is not necessarily always a linear process. It could be a linear process in experiments where we can isolate all other effects and factors. Yet, if we managed to do that in experiments, it is not entirely applicable in real life; it is more complicated than that. The linear effect could still be there; for example, if someone is exposed to mostly happy conditions over a long period and we know the HN- values, we could measure the effects and derivatives that could be referred to as, for example, **HN-emotional stability state.**

The HN-state

"The HN-state" could be another effective tool, where we assign different states on a 1-10 scale. So, for example. HN-state 09 refers to a stable hormonal, neural response over five years as the participant is exposed to mainly the same emotions. A more disturbed state would be, for example, HN-state 04, where the participant is exposed to different emotions and circumstances. The HN-state could have a deeper meaning referring to what we could refer to as the "emotional quantum state." In chapter three, we will discuss a proposed quantum mechanical approach.

In 1980, Robert Plutchik –a psychologist, an adjunct professor at the University of South Florida, and a professor emeritus at the Albert Einstein College of Medicine- diagrammed a wheel of eight emotions: joy, trust, fear, surprise, sadness, disgust, anger, and anticipation. Plutchik also proposed twenty-four Primary, Secondary and Tertiary dyads (feelings composed of two emotions). The wheel of emotions can be paired in four groups: primary dyad, secondary dyad, tertiary dyad, and opposite emotions.

The primary dyad is one petal apart; for example, Love = joy + trust. The secondary dyad is two petals apart, for example, Envy = Sadness + Anger. The tertiary dyad is three petals apart, for example, Shame = Fear + Disgust. Opposite emotions are four petals apart, for example, anticipation $\notin$ Surprise.

There are also triads, emotions formed from three primary emotions. This leads to a combination of 24 dyads and 32 triads, making 56 emotions at one intensity level. Emotions can be mild or intense; for example, distraction is a mild form of surprise, and rage is an intense form of anger.

This is one approach from an experienced psychologist; our target is to validate that and quantize it. There could be many classifications, approaches, and assumptions of which emotions could be "derived" from other emotions or could evolve from other emotions without finding a methodology to quantize and measure that they will remain assumptions.

Measuring the HN-value and the rate of change, and the methodologies proposed in this book could help us in our mission.

How do we deal with non-linear effects and states, then?

The isolation principle proposed earlier in this book would be quite effective here, and generally, the five main principles proposed when quantizing emotions would help. Here is a reminder of these principles:

1. The unique factor.
2. Pre-defined conditions.
3. Imposed conditions.
4. Complex states.
5. Isolation principle.

So, if we apply these principles, we will be able to detect the linear and non-linear effects again using the **HN-value.** We can then isolate the factors and prescribe the necessary required corrective action.

A more complicated approach would be using an equation that tells us in more depth how these emotions evolve in time.

A small reminder on introducing the "in sign to the Schrodinger's equation":

$$i\hbar \frac{\partial}{\partial t}\psi(x,t) = \hat{E} \ast \hat{\psi}(x,t)$$

Here the equation is prescribing that the energy operator is acting "in" a wave function, which is the operatee in this case, and this wave function does not have definite energy; it could be a field of waves. This also prescribes that the energy operator and the wave operatee are affecting each other and can swap roles.

In this case, the analogy would be that we could derive emotions from each other and quantizing how they evolve in time, acting on a field of waves. It is not an easy operation, as we will need to define more basic parameters before we master how to deal with such an equation.

Let`s go one step back to classical mechanics for a particle moving in one dimension; what I need to know about it is the position (x) and momentum (p). Other things like angular momentum, kinetic energy, potential energy, total energy are all functions of x & p.

What happens when I measure any variable for a classical particle? Well, if you know the location, it is guaranteed to be x and momentum to be p; any other function of x & p is guaranteed to have that particular value. Because everything is completely known, that is the situation at one time. Then you want to ask what we can say about the future? What`s the rate of change of these things? And the answer to that is derivatives.

It is a bit more complicated in quantum mechanics. How do you describe the particle in quantum mechanics in one dimension? Will assign to it a function $\psi(x)$ basically, any reasonable function which can be squared and integrated over the real line. Anything you write down is a possible state. That is like saying any x & any p are allowed.

Likewise $\psi(x)$ is nothing special. It can be whatever you like as long as you can square it and integrate it to get a finite answer over all of space, and if all of space goes to infinity, then ψ should vanish at plus & minus infinity. If you have a simple question about our particle, you want to know where it is; that is when you do not get a straight answer. You are then told it can be here, it can be there, it can be anywhere else, and what is referred to as the probability density that it is at point x, P(x) Is proportional to the absolute square of ψ, | $\psi(x)$ |²

That means you take the ψ and square it so it will not have anything negative in it; everything will be real and positive. ψ itself may be complex, but the $\psi * \psi$ is defined to be real and positive.

The analogy here is that it is pretty clear that the classical model will not work quite well for quantizing emotions. One of the reasons is that you do not get completely known values for x & p. In our analogy, the position and momentum of your emotions.

Yet, the quantum mechanical model seems to be fitting our target in quantizing emotions, and following a bit of the methodology of quantum cognition -as explained in this book- if we assign a function $\psi(x)$ to our emotions, we have a wide range of probabilities that we can deal with. What potentially could be different is squaring the function, whether it would be required or not.

Yet, this brings us back to how introducing the "IN sign" to Schrodinger's equation could be helpful here

$$i\hbar \frac{\partial}{\partial t}\psi(x,t) = \hat{E} * \hat{\psi}(x,t)$$

We can use it to study how the "functions of emotions" evolve in time in relation to the energy operators and the field of waves they are basically operating on.

Emotions Energy

Is everything in the universe energy?

We tend to identify energy only through some of its manifestations, such as electric energy, kinetic energy, heat energy, magnetic energy, etc.

And we have four forces: electromagnetic, gravity, weak nuclear, and strong nuclear force.

Another question is that manifestations of energy are limited to manifestations like heat, motion, etc. Or are there other manifestations that could be considered as energy as well?

I will consider this a thought experiment rather than a hypothesis. As mentioned in the introduction, this book does not claim to be proposing a scientific theory. It presents some thought experiments that could open the door for further research and scientific validation.

Let's consider emotions as a manifestation of energy. In other words, let's consider emotions as energy. Earlier in this book, we proposed considering emotions as an energy operator. We will do it in simpler terms and approaches in this thought experiment.

So if someone gets angry, her face turns red. Let's consider that as a form of energy conversion. Let's assign a unit to "Emotions Energy," for example, **EMJ (Emotions Joules)** converted into heat energy leading to a red face. We can then try to measure how the amount of EMJ led to a specific amount of Heat (Joules). We can also consider the rate of Emotions Energy transfer as, for example, **EMW (Emotions watts),** which could be Emotions Joules per second or **EMJ/S**. In this book, we have presented some of the methods used to measure brain activity when responding to emotions. We have proposed measuring it using the **HN-Value** (Hormonal-Neural response) value.

Yet, not every emotion is converted into a red face or heat energy. Some emotions are converted to other "manifestations of energy," which do not fall within our sort of traditional known types of energies.

Some emotions could convert into complex manifestations of energy. As this field of study develops, we will be able to define better and segregate what could be referred to as manifestations of energy due to the conversions of emotions energy. However, till this becomes well developed, we still have a very good tool to measure it through its effects on the body represented in the HN-Value proposed in this book or the hormonal & neural effects of emotions.

Emotions energy could have an effect on an individual; additionally, it could have a collective effect on a group of people, as explained in some of the thought experiments presented in this book, including the football games example and the quantum party thought experiments.

The necessity of motion for emotions energy

Another question arises here: Is motion necessary for our hypothetical emotions energy to affect an individual or a group?

The apparent & sort of intuitive answer would be yes. To have an emotional effect on someone or a group of people, you need to do this effect basically through sound or vision or smell, in other words, through senses. So there should be a motion of sound or light waves, for example.

What we have learned is that the direct answers always have their limitations. In our case, yes, motion could be necessary for the emotions energy in some situations. Still, the question is, could emotions energy have its effects without motion or even without a medium? To answer this question, we will have to borrow some quantum cognition methodologies.

The emotions particle

So we have the Higgs field of the Higgs mechanism explaining why some particles have mass. In the 1960s, Peter Higgs, along with five other scientists, proposed the Higgs mechanism, and they made theoretical predictions for the Higgs Boson, which is an elementary particle produced by the quantum excitations of the Higgs field.

The Higgs mechanism required a spineless particle known as a boson should exist with properties described by the Higgs mechanism theory. This particle was called the Higgs boson. A subatomic particle with the expected properties was discovered in 2012 – More than five decades after the theoretical predictions- by the ATLAS and CMS experiments at the Large Hadron Collider (LHC) at CERN, Switzerland. The new particle was subsequently confirmed to match the Higgs boson's expected properties.

It is worth mentioning that so far, particle physicists cannot directly observe the Higgs Boson; instead, they rely on statistics to know it is there.

Another important theoretical predictions are for the graviton in theories of quantum gravity; the graviton is the hypothetical quantum of gravity, an elementary particle that mediates the force of gravity. In string theory, the graviton is a massless state of a fundamental string. The graviton is expected to be massless because the gravitational force is very long-range and appears to propagate at the speed of light.

Here, we will borrow quantum cognition methodologies to propose considering emotions as a field. Excitations or fluctuations of the emotions field produce the effect and could be referred to as a particle. Comparable to the Higgs boson and the graviton. We could call it the *"Ematon" for example, which will represent the quantum of the field of emotions and could be produced by the quantum excitations of that field.*

This seems like a long way ahead and would require many brains to work on. In the meantime, at least we could make some theoretical predictions, which could start with relevant thought experiments. The field of emotions and a particle of emotion, "Ematon" is at this stage just an analogy. The key idea here is that this analogy could be useful when we want to build more analogies and potentially theoretical predictions giving emotions the properties of particles and taking it from there. Quantum entanglement, quantum coherence & quantum tunneling could be some of the concepts and properties that we can use as analogical models when dealing with emotions again following quantum cognition methodologies.

As explained in this book and in the section "What is a particle?" a particle could be considered as a collapsed wave function or a quantum excitation of a field or a vibrating string. Accordingly, we could analogically consider the emotions particle "Ematon" as a collapsed wave function of a field of waves or a quantum excitation of a field of emotions or a vibrating string (which probably could be massless comparable to the graviton).

If we used statistics to confirm the Higgs Boson's existence, potentially we can build a sophisticated model and theoretical predictions to confirm the existence of an "emotions particle – The Ematon". And as science advances, we could figure out ways to directly observe the "statistically confirmed particles" for the time being and for the purpose of this book. It is still analogies and thought experiments.

Another question arises here following this model, would it be one particle for emotion, or could we develop a "Standard Model of Emotions" where each specific emotion has its field and, respectively, its particle. We will discuss this in a later section in this book.

Quantum entanglement of emotions

So back to the question, is motion necessary for emotions energy to have an effect? And could emotions energy have its effects without motion or even without a medium?

Quantum entanglement occurs when two particles become connected in such a way that when the property of one particle is changed, an instantaneous change in the property of the other particle occurs. Entangled particles have opposite properties or states. Particles have a property called spin, and the particle will either be spin up or spin down in any given direction. When two particles are entangled, and their spins are measured in the same direction, one particle will be spin up, and the other particle will be spin down.

According to quantum mechanics, a pair of entangled particles could be separated by even an entire universe. When the state of one is measured, and its superposition collapsed, it would immediately collapse the superposition of the other particle. Measurement would indicate that it had the opposite state of its entangled partner. This seems to say that entangled particles can communicate at a speed faster than the speed of light. Something that Einstein`s theory of special relativity ruled out. Einstein famously called this phenomenon spooky action at a distance.

Following quantum cognition methodologies, could emotions be entangled? To test that probably we would need to define the field of emotions; we would need to discover the emotions particle even statistically. Then we could test the entanglement of the emotions particle. Yet this full model or proposal is based on a quantum cognition approach. As this field of study develops, there could be other ways to approach the idea and test it.

This, too, seems like a long way ahead and would require many brains to work on. In the meantime, at least we could make some theoretical predictions, which could start with relevant thought experiments.

The idea of the quantum entanglement of emotions gives us an insight that "emotions energy" does not necessarily need motion or even a medium to make an effect.

One example where a motion and a medium play a role in the equation is when sound waves affect an individual or a group. Another example is when a natural view affects an individual or a group.

However, it is not merely the sound waves causing the effect. The information carried within these sound waves and the way they are perceived by the individual or the group of people are playing an important role. This brings us back to the concept introduced in this book of swapping roles between the operator and the operate. A reminder of the equation proposed in this book and explaining the idea:

$$\hat{E} \divideontimes \hat{s} = f\Psi(x,t)$$

Please refer back to Energy in shape = Function thought experiment for further details and explanation.

Do we need a periodic table & a standard model for emotions?

Let's start with the periodic table of the chemical elements. At the end of the 19th century, the understood theory of what the universe is made from is that there are more than a hundred different chemical elements. And thanks to Dalton's atomic theory, what it said is that for every element, hydrogen, helium, lithium, etc., there is an atom which is a fundamental, indivisible, indestructible little thing, and you had different atoms, one for every element. So, that was a sort of Victorian view of the nature of matter.

At the turn of the 20th century, 1897, a particle was discovered. The first elementary particle that we found was called the electron, which led over the next few years to a revision of the atoms' structure. The atom was thought to be something hard, indestructible, and indivisible. When the electron was discovered, that was revised. And we get the model of the atom that we currently learn about in school, which is a nucleus that contains most of the mass of the atom and that's positively charged, and around the atom go these electrons.

Now, if you look at the periodic table, the way that the elements were arranged, there are certain patterns in the properties of the different chemical elements. So, for example, if you look at the group 1 elements, they all tend to react in similar ways, and they get more reactive as you go down. There are clear patterns in the way the elements are arranged. That was sort of indicative of some deeper structure. So, essentially you can explain the properties of all these different elements by different numbers of electrons going around the outside of atoms, and those electrons are what determine the chemical properties of that particular element.

This is not the end of the story; in the 1930s, it was discovered that the nucleus itself is made of smaller things called protons and neutrons. These are smaller particles that make up most of the mass of the atom. The proton is positively charged, the neutron is electrically neutral. They are much heavier than the electrons; they are about 2,000 times more massive than electrons.

This may be where your school physics ended possibly, but in the 1960s that the protons and neutrons themselves are not fundamental; they are made of smaller things called quarks. The proton is made of two up quarks and one down quark. The neutron is made of two down quarks and one up quark. That says that all of matter, every atom in the universe, everything that we know about is basically made of just three different elementary particles, the electron and the two quarks (the up quark and the down quark). So everything that exists is made of just these three things. So you are just quarks and electrons arranged in a rather peculiar way. And these are the first three particles of what we call the Standard Model.

Some consider the Standard Model of particle physics as the closest we have so far for a sort of complete description at the fundamental level. So we got the electron, the up quark, and the down quark, and then there is something else that gets added to this table called a neutrino. Neutrons are sort of like ghosts; they are invisible, almost undetectable particles. There are trillions of them going through you right now. They are produced by the sun in vast quantities; they go straight through you, straight through the earth, and they very rarely interact with the ordinary matter we are made out of. That's why we are not aware of neutrinos' existence most of the time.

The column of electrons, up quarks, down quarks, and neutrinos make up what we call the first generation of matter. For some reason which is not fully understood yet, nature provided us with two additional copies of these particles. In the second generation, all the particles are exactly the same as in the first generation, except they are more massive and they are unstable. So, for example, the electron has a sort of a heavy cousin called the muon, which is about 200 times more massive than the electron. And the reason they were not made out of muons and that there are not muons hanging around is that if you make a muon, it will very quickly decay into an electron and some neutrinos. So these second-generation particles do not hang around very long. They are unstable, but you can make them in high-energy collisions like at the LHC, for example.

Then there is a third-generation, which is even heavier. The three generations are the 12 matter particles. They make up the kind of the solid stuff of the universe, or at least they would if they weren't all unstable apart from the first column. And it is a big mystery as we do not know why there are two extra columns of the second & the third generation in the Standard Model of particle physics table. It is a bit like the periodic table in a way where you have this sort of structure, and you can see these patterns. But you do not understand yet – back in the 19th century- what underlies this. But there is something suggestive here, something that sort of hints that may be that there is some deeper structure why we have got this rather peculiar set of matter particles. Before we come to that, let's see the last ingredient of the Standard Model of particle physics is the Force Particles.

There are three fundamental forces in the Standard Model, Electromagnetism (That's the force that causes electrons to stick to the nuclei of atoms, it binds atoms together.) And the particle that transmits the electromagnetic interaction is the photon – the particle of light). So light itself is also an electromagnetic phenomenon.

Then there is the gluon, which is the force particle for the strong nuclear force, which is a force that binds quarks together inside the atomic nucleus and binds protons and neutrons together. It is called a gluon because it essentially glues things. And then there are the W & Z particles, and these are particles that transmit the third force, which is the weak nuclear force. The weak nuclear force does not bind things together like electromagnetism or the strong nuclear force. The weak nuclear force is responsible for causing particles to decay. So when a muon turns into an electron, that happens through the weak force.

This was the Standard Model of particle physics as it had been studied and observed until the 3rd of July 2012; there was one missing piece, the Higgs. We gave a brief about the Higgs Boson earlier in this book. Yet, we need to mention that what we think of as being fundamental are not particles but fields. If you have ever held a magnet near to a piece of steel or iron, you felt the effect of a field. An example of a field is a magnetic field that can be strong near a magnet and gets weaker as you move further away. Or it could be a gravitational field, like the one that the earth creates around it or the sun creates around it.

In particle physics, every particle in the Standard Model has an associated field. So, there is a field for the quarks, a field for the electrons, a field for the neutrons, and even for all the force particles. One way to think of these particles is like little tiny ripples moving through these fields, which we explained in more detail in the "What is a particle?" section in this book.

And that`s the way particle physicists generally tend to think of all matter, so electrons, quarks, everything is just little ripples moving through this cosmic energy field that fills all of the space and is everywhere.

The way the periodic table and the Standard Model sort of evolved are very powerful and could help develop a similar methodology when quantizing emotions. We do not have a "periodic table for emotions" yet. Some comparable attempts are the "Wheel of emotions" diagramed by Robert Plutchik in the 1980s and mentioned in this book in the "derivatives of emotions."

Yet, we need to systematically develop a "periodic table for emotions" to understand the patterns that this periodic table could tell us. Once these patterns are well understood and studied, this could lead to further and deeper findings on quantizing emotions.

The story does not end here as well; we do not need centuries to conclude that we can study emotions as a field. Happiness, sadness, fear, etc., are sort of the ripples moving through this cosmic energy emotions fields that fill all of the space and are everywhere. In a very comparable way to how we are currently studying the Standard Model of particle physics, we can develop the Standard Model of emotions or the Standard Model of Qualitative Physics.

As mentioned earlier in this book, we do have three core effects as most researchers would agree on what is usually referred to as the basic or primary emotions to be six: Happiness, sadness, fear, anger, disgust, and surprise, which are differentially associated with three core affects reward (happiness), punishments (sadness), and stress (fear and anger).

Yet, to properly conclude whether the three core affects reward (happiness), punishments (sadness), and stress (fear and anger) are sort of the fundamental particles of the proposed Standard Model of emotions or the Standard Model of Qualitative Physics, this requires further research, experimentation, and validation.

Developing a "periodic table of emotions" and a Standard Model of emotions or the Standard Model of Qualitative Physics even analogically would assist in our attempt to quantize emotions.

Though a periodic table of emotions would seem to be reverse engineering, we would expect that the Standard Model is a better fundamental description. Yet, a periodic table of emotions could help categorize, observe specific patterns and potentially other conclusions.

Could it be more than one field, not just a field of emotions?

In the previous analogies used in this book, we have mentioned the field of emotions on few occasions. However, could we use a different analogy and a different approach? Could we consider that each of the "fundamental emotions" has its own field? For example, if we consider three fundamental emotions happiness, sadness, and stress (fear and anger), each having its own field. For example, let's consider a field of happiness and its effect on a specific person is the ripples or fluctuations of this field in this specific person leading to the hormonal and neural response.

We can use this analogy in the case of a collective feeling of a city's residents or even a full country where they have a general feeling of happiness or stress. It is analogous to a field of this specific emotion in this city or country. The ripples could be emphasized or get stronger as this collective field gets emphasized, comparable to the vibrations of the strings getting steeper as a result of stronger resonance, for example. There is an interesting field called Urban Psychology studying the psychological effects of cities on their inhabitants and how they affect the ecosystem and get psychologically affected by the way the city is built and planned.

Crowd psychology, also known as mob psychology, is a branch of social psychology. Social psychologists have developed several theories for explaining the ways in which the psychology of a crowd differs from and interacts with that of the individuals within it. Major theorists in crowd psychology include Gustave Le Bon, Gabriel Tarde, Sigmund Freud, and Steve Reicher. This field relates to the behaviors and thought processes of both the individual crowd members and the crowd as an entity. Crowd behavior is heavily influenced by the loss of responsibility of the individual and the impression of the universality of behavior, both of which increase with crowd size.

The football game is another analogy where a field of a specific emotion like the happiness of the winning team and anger for the losing team is sort of propagating through the crowd, and its ripples are sort of measurable in the individuals.

If we move on with the analogy of the Standard Model of emotions or Standard Model of Qualitative Physics, the model would suggest that at least each of the fundamental emotions would have its field. This field's ripples could be considered the particle of this specific emotion.

If you follow the logic of the model, an inevitable question will seem to come up: Would we need sort of a "Higgs-like emotions particle" or rather field that would give other emotions their "mass" or meaning. In this book, I have proposed the "emotions particle – The Ematon" which could be a good candidate. The Ematon could be a "Higgs-like emotions particle" or field that would give other emotions their "mass" or meaning. And the other fundamental emotions would have their associated fields and particle.

Another field and associated particle which would fit quite well in the analogy is a "Gluon-like emotions particle." Maybe I have gone so far with the analogy. But since we have gone down that road, let`s finish it to the end and see how far the analogy would go.

Let`s consider Force-like particles for emotions in the Standard Model of emotions.

We would need an Electromagnetism-like force, and we would need a photon-like particle that transmits the interactions of emotions. Light does affect our emotions, but could it be having the same function of transmitting the interactions of emotions? Or do we need to search for a better candidate?

The gluon is the force particle for the strong nuclear force, which is a force that binds quarks together inside the atomic nucleus and binds protons and neutrons together. It is called a gluon because it essentially glues things. So our "Gluon-like emotions particle "could potentially be binding other fundamental emotions together.

Then, the W & Z particles transmit the third force, the weak nuclear force. The weak nuclear force is responsible for causing particles to decay. Accordingly, our "W & Z- like emotions particle" could potentially be responsible for the decay of the particles of emotions.

Excuse me if I went so far with the analogy. It could either turn out to have some promising predictions, it could inspire some other ideas and predictions, or at least it could ignite your imagination a bit while reading on the beach or something.

Again it is just an analogy or sort of a thought experiment and not claiming to be proposing any sort of scientific theory. Yes, the analogy is built on very solid and well-recognized physical theories and then borrowing these concepts for analogies and thought experiments to understand better and quantize emotions. This methodology has been proven successful when used in quantum cognition, as explained in this book's quantum cognition section. However, again, it`s merely proposing analogous models for further discussions and validations.

Yet, if we want to test some of these ideas in a more quantifiable way, our best shot, for the time being, would probably be building statistical models and designing experiments accordingly. Potentially using the HN-values proposed in this book would help as well.

What is a particle?

In the late nineteenth century and with the discovery of the electron, we knew that they are negatively charged "particles." This was based on the work of the scientists during the 1880s and '90s when they searched cathode rays for the carrier of the electrical properties in matter; their work culminated in the discovery of the electron in 1897 by the physicist J.J. Thomson. It only took us less than three decades to discover the electron-possessed spin. With progress being made in physics generally and quantum mechanics, specifically, our understanding of "particles," including electrons, started developing.

If you ask a group of particle physicists what a particle is, you might get different answers on the definition of the particle. One of the interesting definitions is that a particle is "a collapsed wave function." The wave function representing an electron, say, is spatially spread out so that the electron has possible locations rather than a definite one, but when you measure the electron's location using a detector, its wave function collapses to a point, and the particle clicks at that position in the detector. Making a measurement is establishing an entanglement between the detector/device and a system.

Another interesting definition of the particle would be "A particle is a quantum excitation of a field" Quantum field theory is the math where particle physics is written. In that, there are a bunch of different fields, each field has different properties and excitations, and they are different depending on the properties, and those excitations we can think of as a particle. As physicists discovered more of nature's particles and their associated fields, a parallel perspective developed. The properties of these particles and fields appeared to follow numerical patterns. By extending these patterns, physicists were able to predict the existence of more particles. "Once you encode the patterns you observe into the mathematics, the mathematics is predictive; it tells you more things you might observe," explained Helen Quinn, emeritus particle physic at Stanford University. The patterns also suggested a more abstract and potentially deeper perspective on what particles actually are.

"What we think of as elementary particles, instead they might be vibrating strings" – Mary Gaillard. Physicists are trying to fit symmetries inside a single, larger group of transformations. The idea is that particles were representations of a single symmetry group at the beginning of the universe. Researchers placed high hopes in string theory: the idea that if you zoomed in enough on particles, you would see not points but one-dimensional vibrating strings. You would also see six extra spatial dimensions, which string theory says are curled up at every point in our familiar 4D space-time fabric. The small dimensions' geometry determines the properties of strings and, thus, the macroscopic world. "Internal" symmetries of particles, like the SU(3) operations that transform quarks` color, obtain physical meaning: These operations map, in the string picture, onto rotations in the small spatial dimensions, just as spin reflects rotations in the large dimensions. "Geometry gives you symmetry gives you particles, and all of this goes together," Nanopoulos said.

The theoretical physicist David Tong explained is that there are fields that underlie everything. And what we think of as particles are not really particles at all; they are waves of these fields tied up into little bundles of energy. There are no particles in the world; our universe's basic fundamental building blocks are these fluid-like substances that we call fields.

Quantum cognition

Some psychologists think that a set of ideas borrowed from quantum mechanics could help us make sense of human behavior. The notion is called quantum cognition, suggesting that quantum mechanics' mathematical tools could help make human behavior more predictable. Quantum mechanics is all about statistics; for instance, it is impossible to know where an electron is at any given time, and you can only know how likely it is that you will find it at a given place when you measure it. Statistics are also useful in other branches of science for different reasons; they can help us understand the big picture even if we don't know all the lower-level details. And when it comes to the brain, there are a bunch of low-level details that we don't understand. But in some cases, we can use statistics to make decent predictions about how people will behave even if we are not really sure why. Traditionally, these cognitive models have relied on classical probability. Human cognition is full of ambiguities, and classical probability is not well-suited to handling them. That is where the quantum cognition notion considers quantum mechanics to be useful.

Recently, psychologists have been exploring whether or not quantum mechanics tools could be reused to help understand the brain and predict human behavior. For example, it is really tricky to predict how humans will make decisions. The things humans decide just don't always seem "logical." Classical models of cognition struggle to explain it, principles from quantum mechanics help to understand it. In quantum mechanics, not knowing could produce an unexpected result, which is comparable to the human decision-making process; not knowing or not being sure of the results could produce unexpected decisions. As illogical as these scenarios sound though neither one is unpredictable. Quantum probability theory is a model that accounts for the fact that knowledge of something can affect it. The same theory could be applied to human decision-making to predict how people will make decisions, even we do not understand precisely why. Psychologists were able to use this quantum model to correctly predict people's decisions in a coin flip experiment even when their classical models failed.

In general, the way someone answers a question can be unpredictable for a lot of reasons. In fact, just the order of questions you ask someone has a big effect on the answers they give. For example, imagine I ask you how your vacation was? And after you answer, I follow that up with, "How did you get along with your boss? But if I asked those questions in the opposite order and if you'd gotten a fight with your boss, you might say that you have liked your vacation less than you would have otherwise. The order of the question can change your answer, and classical models have trouble explaining that, but quantum mechanics might be able to help here too. In quantum mechanics, things that seem like basic math like multiplication are not so straightforward; for instance, A times B will often give you a different answer than B times A. So quantum models have to account for those rules to make predictions. Psychologists have managed to do a similar thing. Using quantum-inspired math, psychologists built rules that account for order into their models. Because of that, they were able to make a specific numeric prediction about how an experiment would turn out before it even happened, which is rare in psychology.

Psychologists often don't know enough about the underlying causes of a person's behavior to make specific predictions. But in this experiment, a team of researchers analyzed 70 national surveys in the US conducted by Pew & Gallup that randomly ordered pairs of questions. They made a specific prediction about how the questions' order would affect the answers, and the results proved they were right. This suggests there are ways to predict human behavior, even when it does not seem to make logical sense.

Human behavior is not the only element of our cognition that can be unpredictable. Even the things we perceive can sometimes seem inexplicable. Bistable perception or (Multistable perception) is an example of that, which is a perceptual phenomenon in which an observer experiences an unpredictable sequence of spontaneous subjective changes. The tools of quantum mechanics can give us a way to understand it. For example, in 2007, a team of scientists modeled the brain's response to the Necker cube as a simple, two-state quantum system. The system modeled how often a person's view of the cube flipped from one version to the other. Then the scientists compared the frequency of those flips to two biological time scales. The amount of time it takes the brain to process new sensory input and the amount of time it takes us to react to that new information. When they compared this quantum-inspired model to past experiments' real results, they found that it generally predicted the way people`s brains responded to this illusion.

Roger Penrose (an English mathematical physicist, philosopher of science, and Nobel laureate in physics) published a book in 1989 titled "The emperor`s new mind," where he advocated that classical physics is not adequate to explain consciousness. The brain harnesses quantum mechanics and wave function collapse. He discussed the possibility of quantum processing in the brain.

Yet, quantum cognition psychologists' main focus is building models on principles drawn from quantum mechanics. This model addresses puzzling cognitive phenomena. These quantum models show promise in addressing cognitive phenomena that have proven challenging to be modeled by means of classical probability theory. These models have shown significant predictive ability.

Quantum biology is another interesting field of science studying the applications of quantum mechanics and theoretical chemistry to biological objects and problems.

So, is what I am proposing in this book considered quantum cognition?

The mathematical tools of quantum mechanics could help make human behavior more predictable. Human cognition is full of ambiguities, and classical probability is not well-suited to handling them. That is where the quantum cognition notion considers quantum mechanics to be useful. Psychologists were able to use this quantum model to predict people's decisions correctly. These quantum models are promising in addressing cognitive phenomena.

In this book, quantum cognition methodologies have been borrowed repeatedly to help drive analogies, thought experiments, and approaches in an attempt to quantize emotions.

For example, in this book, I have proposed applying the "IN sign" to Schrodinger's equation, and potentially it could be introduced to other equations in physics and mathematics. Schrodinger's equation itself integrated with the "IN sign" could help understand how emotions evolve in time the same way it is describing how the wave function evolves in time. Whether or not this could be considered quantum cognition or categorized differently, the main goal is to define a reliable, valid measuring methodology for quantizing emotions.

Chapter 04

Evolution of emotions

A bit about the evolution of emotions

The evolution of emotions is something that Charles Darwin himself explored". Darwin is seen as the founder of the evolutionary approach to emotions. He said that emotions were evolutionarily evolved and that they were here to serve an important survival purpose. He actually argued that emotions have evolved via natural selection and serve this purpose for humans' survival and helping aid in communications and bonds with others. For Darwin, emotion had an evolutionary history that could be traced across cultures and species.

In 1872, he published his landmark book, "The Expression of the Emotions in Man and Animals." This book was immediately the best seller of its time, over 5,000 copies, which was a lot in those days. It was considered the first scientific treatment of emotion. It included survey data, observations of both healthy and mentally ill individuals looking at healthy and unhealthy emotions.

Darwin analyzed emotional responses in animals. He compared such responses in a variety of species and proposed that expressions of emotion evolved in an organism so as to signal to other organisms what it is about to do next. For example, when an animal is feeling threatened, it may display a stance of aggression. It may bear its teeth, it may lean forward, and it may make aggressive sounds. All of this behavior is to indicate violent behavior, and it does well in intimidating some percentage of other organisms nearby, which may then retreat in fear without necessitating actual combat. Thus this type of behavior has a quantifiable impact on the organism's survival and therefore becomes a preferable trait within the gene pool.

To give a brief about how Darwin approached the evolution of emotions, we can think of three main principles that he talked about to help explain the origins of emotions:

1. The principle of serviceable (or useful) habit.
2. Principle of antithesis.
3. Principle of involuntariness (Nervous discharge).

The first principle: The principle of serviceable (or useful) habit which is very important in understanding how emotions have evolved. Darwin argued that useful habits become reinforced over time. For example, think of the emotion of disgust. Disgust is pretty useful in helping deter us away from ingesting substances or toxins that might make us feel ill. So we have disgust as an emotion because it serves as a useful or serviceable habit for us.

The second principle: Principle of antithesis. While some emotional habits are useful, like the case of disgust, other emotions are what Darwin called emotional habits to occur merely because they are their opposite or antithesis. For example, if you encounter some opposite situation in which you typically elicit a habitual response, rotten food, for instance, an opposite manner of the habit will occur, such as the opposite of the emotion of disgust. This principle was the one that received perhaps the least attention of all three.

The third principle: Principle of involuntariness (Nervous discharge). Where some involved behaviors or habits are performed as the result of built-up energy in the nervous system as and this will lead to nervous discharge, as Darwin put it, for example, when you are in an extreme situation of threat where you feel fear, he would say you will have what he called a nervous discharge of screaming. In more modern-day situations like for example, you are sitting with a boss or someone that you are very uncomfortable around, and your nervous discharge might just be even your leg shaking or your arms trembling. So he would say this is the third principle where some evolved behaviors or habits are performed as a result of this built-up energy in our nervous system.

Darwin considered most of our emotional expressions are innate, but he did argue as well that not all emotional expressions are sort of instinctive, that there are some things which would say like kissing, for example, that are learned but are used to fulfill a more evolutionarily primal desire. So most of our expressions are instinctive, but those that are not are things we learn to fulfill more evolutionarily or instinctive needs.

The EEA or the environment of evolutionary adaptiveness is another important concept in understanding emotions' evolution. The EEA is the environment that forms sort of the origins of humankind, and it is where emotions are thought to have first evolved.

So what is an accurate picture of the environment in which human emotions are first developed or originated?

In an attempt to answer this question, researchers in this field of study often try to look into other pre-industrial societies like hunter-gatherers societies to think about what might be close analogies that might help explain some of our early emotional heritage. Some of the EEA properties are:

1.   Vulnerability of offspring.

2.   Monogamous bonds.

3.   Emergence of caring and compassion.

4.   Flattening of social hierarchy.

5.   Need for collective action.

The EEA has many properties that seem to give rise to social emotions ranging from compassion and love to cooperative behaviors and feelings. The EEA also resulted in some selection pressures that favored certain genes more than others, and these genes drive which emotions to become adaptive and universal across all human beings.

Vulnerability of offspring: After infants are born, they are highly dependent on their caregiver for a long period of time before they can function independently. So because as humans we are very dependent on our caregivers for many years, it is really important to have emotions that are sort of built-into us that help us want to care for our young and offspring; in other words, emotions like compassion or pro-social feelings of nurture and love became really important in our early history as a vital part of promoting this long term care of off-spring in hunter-gatherers societies. So what seems to be the case is that EEA would be the kind of environment where compassion and love would be really important emotions.

The second property which is Monogamous bonds. Human society seems to have a strong underlying tone of monogamy or long-term commitment to a partner to help rear this vulnerable offspring. An interesting comparison that suggests the evolutionary nature of emotions is that the monogamous bonds for humans are actually in contrast to bonobos, such as having sex often and with many other monkeys. Before you start denying the evolutionary nature of emotions because it did not select this specific trait for us, bonobos often do this to ease aggression; any conflict in the bonobo society is addressed with sexual resolution. Yet this behavior did not seem to save them from being an endangered species. For humans, that is often less the case; maybe for the most part, at least for rearing offspring, this monogamous is really important. So we need to have emotions sort of built-in us that enable us to be connected in a long term way to a partner to help care and sort of rear these offspring.

The third property which is the Emergence of caring and compassion. Part of this is because hominids live longer, so there were older people who also required caring. So it is both the existence of vulnerable offspring, infants and young children, as well as older people who have led to societies where it is really important to take care of others who are vulnerable, so these emotions of caring for others and having compassion is a vital part of our early origins.

The fourth property of the EEA, Flattening of social hierarchy. Researchers suggest that the social hierarchy in hunter-gatherers societies is fairly egalitarian. We see hominid groups of 75 to 100 individuals are pretty egalitarian in a sense. Jed Diamond, a researcher on the topic, argued that it is not until more recently, with the development of agriculture and excess food production, that has led us away from our egalitarian roots.

The fifth property of the EEA which is Need for collective action. As a result of living in these large egalitarian groups, a central and important part of society includes cooperating with others.

Researchers Leda Cosmides and John Toby from UC Santa Barbara, who focus on the evolutionary approach to emotion, mentioned that "Emotion is the statistical composite of selection pressures that caused the genes underlying design of an adaptation to increase in frequency; until they became species-typical or stably persistent."

In other words, they thought of emotions as hardwired programs or modules. The mind is composed of various modules or mini-programs to solve recurring evolutionary issues. Programs are "turned-on/triggered" by certain elicitors in our environment.

Paul Ekman listed six features of what an emotion is from an evolutionary perspective:

1. Brief.

2. Unbidden.

3. Cross-species.

4. Coherent.

5. Fast.

6. Automatic appraisal.

Psychoevolutionary Theory (PET) asserts that humans' positive reaction in nature is programmed evolutionarily. Nature was the first place humans learned to survive by gathering the resources around us, just as our evolutionary ancestors did. Because of evolution demands, which mostly include survival and reproduction, humans have built-in positive reactions to natural environments. PET explains how being in nature can reduce stress and improve your mood but doesn't fully explain how being in nature improves cognitive performance. Can nature make you smarter?

Evolutionary approaches to emotions have traditionally focused on a subset of shared emotions with other species, characterized by distinct signals, and designed to solve a few key adaptive problems. An evolutionary psychological approach broadens the range of adaptive problems emotions have evolved to solve. Additionally, it includes emotions that lack distinctive signals and appears to be unique to humans, and it synthesizes an evolutionary approach with an information-processing perspective.

A number of pathways carry signals from the amygdala to structures in the brainstem that control the emotional response, like the hypothalamus, which directs the sympathetic response, and the periaqueductal gray matter of the midbrain that directs the behavioral response. The amygdala is not a singular structure; it is a cluster of nuclei, which is why sometimes we refer to it as the amygdaloid complex. Each of these regions serves a different function, as has been evidenced by studies on fear, demonstrating that the lateral nucleus is involved with conditioned fear, while the central nucleus controls defensive behavior and so forth.

There are three kinds of major divisions of the brain that are thought to be included in emotion, so this included first are the limbic system, the hypothalamus, amygdala, and nucleus accumbens, and several other nearby areas. The second is the Striatal System or Striatum, which includes the putamen, the caudate nucleus, nucleus accumbens, and it is divided into both the dorsal and ventral striatum. This is often thought of as the reward center of the brain, so it's often associated with the experience of pleasure or reward. The third is the neocortex;

this is the outer layer of the brain. The prefrontal cortex is in the front part of the brain, and it`s often thought of as the prefrontal cognitive control region is thought to be involved in the regulation or control of emotion.

From the standpoint of cognitive neuroscience, there is still much work to be done surrounding the brain mechanism of human emotion. It appears that the brain activity associated with each individual emotion is not highly localized, but it is rather diffuse. Each emotion or even empathy experience when observing others' emotional states is accompanied by activity in the motor cortex and sensory cortex. Brain imaging studies show that similar brain activity patterns are enacted whether a person experiences an emotion, imagines that emotion, or observes someone else experiencing that emotion.

The future of evolution of emotions

A dog story

Let me tell a story about dogs, and our good friends are giving us exciting inspirations on the evolution of emotions. We have many interesting insights from the psychoevolutionary theory of emotions, but I will give you a break from theories for now, and let`s talk about dogs.

Modern dogs belong to the subspecies known as Canis Lupus Familiaris. We can trace their origins back to a now-extinct species of wolves from the Pleistocene, an ancestor they share with the modern gray wolf called Canis Lupus but the exact species of this ancestor is still unknown. From studies of dogs and wolves' genomes, we can say that wolves and dogs began to genetically diverge from each other sometime between 40,000 and 27,000 years ago. Dogs still looked pretty wolf-like at the start of domestication.

These two species diverging genetically is not necessarily the same thing as domestication. One is just a split in the gene pool, while the other is what we are looking for here, and it is the whole behavioral and genetic process that humans were involved in. How do dogs' behaviors and emotions evolved to be what we experience now as our good, loyal and friendly companions?

One of the key genetic traits wolves and modern dogs share – that has been really strongly selected for in modern dogs – seems to be hyper sociability, which is the tendency for adult animals to initiate social contacts even with members of other species. And for some wolves, this tendency along with other behaviors, like scavenging for food, could have made them a better fit for eventual domestication.

So how did the story begin and sort of evolved?

These traits would have been useful as human settlements became more widespread with resources that these canines definitely would have wanted. This is known as the commensal pathway to domestication, where an animal benefits from a relationship with a human, but there is little to no benefit for the humans themselves, well, at least at first.

In our story, proto-dogs were drawn to the discarded human food, which also likely attracted other animals that they could have preyed on, too, and there seems to be some evidence that this was probably happening around 28,500 years ago. They did not have any idea that after tens of thousands of years, humans will be going to the super-markets to buy good food for their descendants.

Eventually, humans realized that wolves could be useful; once domesticated, they could be guards, work with hunters, and even help domesticate other livestock species. After that, whenever humans went, their canine companions followed.

We can track the spread of agriculture through a particular genetic adaptation in dogs! In 2013 scientists were able to isolate the gene associated with the change from the carnivore's diet of wolves to a more starchy diet in dogs. Domestic dogs have more copies of the gene known as AMY2B than wolves do. AMY2B codes for an enzyme secreted by the pancreas to break down starch. An increase in starch consumption in people is often associated with agriculture, like growing wheat and rice. And domesticated dogs living in human settlements would've been fed the kinds of things that people were eating too.

While we are still figuring out when all of this domestication took place and how it happened, it did not seem to take that long before people were deeply attached to their pups, and we can see that bond in the archeological record with burials.

Over thousands of years, domestication led to physical, genetic, and behavioral changes in dogs. While many early dogs looked pretty similar to each other, new breeds were developed to meet various human needs, and coat colors and textures became more diverse. Many of these changes can be traced to the cross-breeding and hybridization of the individual dog population as humans moved around the planet with their canine companions and came across new groups of canids. Today there are hundreds of dog breeds, and most of them are not that old.

So dogs were originally drawn to our ancestors for food, and they eventually bonded with us, working and living alongside us for thousands of years. But the origins of this relationship are still more complicated than scientists originally thought, with discoveries changing the history of dog domestication all the time.

It is interesting how dogs' behaviors and emotions have evolved to be less aggressive and submissive. A trait of hyper sociability has evolved to super loyal and friendly behavior. Some aggressive and hostile behaviors have been either almost gone or channeled.

Is not violence and war driven by aggressive and hostile human behavior an ugly legacy of our evolutionary process? And as we evolve, should not we get rid of such ugly behaviors. We should learn from dogs and evolve beyond aggressive and hostile behavior to more towards friendly and peaceful behaviors. We should evolve to realize that it is way smarter to build schools, hospitals, and research centers rather than weapons factories and atomic bombs.

We have sort of tuned down some of our aggressive behavior to some extent. For example, centuries of war between England and France have now turned into mostly peaceful football and sports rivalry. Well, there are very few acts of violence and hooligans here and there, but this cannot be compared to centuries of war and mass violence. So, we are kind of on the way to becoming more friendly and peaceful, but we definitely need to fast-track it.

This tuning down of aggressive and violent behavior could be noticed in the example of the Roman Empire's famous gladiators. Gladiators were armed combatants who entertained audiences in violent and deadly confrontations with other gladiators, wild animals, and condemned criminals. The Colosseum in Rome was a famous arena for such violent and deadly practices. It is good news for humanity that we managed to find more peaceful and less violent entertainment. The sound of football fans cheering at the modern Stadio Olympico in Rome is way better and more civilized than the sounds of the crowd cheering for the bloody killing of one of the gladiators in the colosseum thousands of years ago.

Back to our dog story, researches show that several groups of genes in humans and dogs have been evolving in parallel for thousands of year including those related to diet and digestion, neurological process and disease. Humans' and dogs' shared environments likely drove this parallel evolution.

Research studying the genome sequencing of humans and canines found that sequences for things such as the transport of neurotransmitters like serotonin, cholesterol processing, and cancer have been selected for in both humans and dogs. Besides sharing genes that deal with diet and behavior, dogs and humans also share diseases, including obesity, obsessive-compulsive disorder, epilepsy, and some cancers, including breast cancer.

Tracking the epigenetic changes that led to this behavior evolution in dogs would be challenging. However, tracking the genetic changes would be easier to do with our available current tools. Additionally, having a mathematical tool like the HN-value (Hormonal-neural response) proposed in this book would also help build models that we can replicate and fast track the evolution of aggressive and hostile behavior towards friendly and peaceful behavior. As suggested earlier in this book, as this field of study develops, we should be able to stimulate epigenetic changes that could lead to this kind of favorable evolution of behavior, which should be a more ethical and sort of organic practice.

Before we wrap up our dog story here, let`s make an interesting comparison between the evolution of dogs and cats' behavior. Cats followed the same domestication path as dogs. The commensal pathway to domestication. Given that for thousands of years, they stayed close to human dwellings for food. Eventually, people noticed that cats were actually pretty good at catching the pests that plagued their food stores and began to entice them to live in their settlements actively.

Today`s domesticated cats are their own species, known as Felis Catus. And we can trace its origins to a species of wild cats called Felis Silvestris, which is made up of five different subspecies. Based on studies of modern house cats' genomes, one subspecies, Felis Silvestris lybica is the direct ancestor of all domesticated cats. These wild kitties are solitary creatures; they do not have the same social structure that other animals like wolves do.

So scientists think that cats' domestication was probably a different process from other animals' domestication. A species is considered domesticated when it becomes genetically and permanently modified through human-influenced breeding. It has to be reliant on humans on some level, like for food and shelter.

Domestication as a process has made hose cats smaller than their wild cats' ancestors and resulted in new varieties in coat color and patterning. These included new variations of the tabby coat. Most of these coat changes are relatively recent as recessive genes in wild cats became more prominent. By the 19th century, researchers suggest that people started to breed for more variations in markings and colors selectively. But beyond size and color, domestication did not really change cats' morphology that much compared to dogs, for example, which have seen significant changes to their whole bodies. This is mostly because of differences in breeding practices as different dog breeds were bred for specific purposes.

Modern cats also maintain more genetic and behavioral similarities with their wild ancestors than most other domesticated animals, including eating and breeding behaviors. This is probably because of interbreeding between domesticated cats and the surrounding wild cat population.

Does natural selection apply to emotions, or are emotions a driving factor for natural selection?

Natural selection is the process that results in the adaption of an organism to its environment using selectively reproducing changes in its genotype or genetic constitution. Those variations in the genotype that increase an organism`s chances of survival and procreation are preserved and multiplied from one generation to another at the expense of less advantageous variations. Evolution often occurs as a consequence of this process.

Natural selection may arise from differences in survival, infertility, rate of development, mating success, or any other life cycle aspect. Natural selection multiplies the incidence of beneficial mutations over generations and eliminates harmful ones. It enhances the preservation of a group of organisms that are best adjusted to their environment's physical and biological conditions and may also result in their improvement in some cases.

So has emotions evolved via natural selection?

Darwin argued that emotions have evolved via natural selection and serve this purpose for humans' survival and helping aid in communications and bonds with others. For Darwin, emotion had an evolutionary history that could be traced across cultures and species.

Let`s use an analogy of applying the natural selection mechanism to emotions by going back to our dog story. A significant behavioral trait in our modern dogs` ancestors, hyper sociability, evolved, leading to more beneficial and enhanced behaviors that led to better-surviving conditions for the species and significant improvements and enhancements. Physical gains and kind of lifestyle improvements, so after roaming human settlements for food leftovers tens of thousands of years ago, their descendants now have dog hotels, care centers, clinics, and premium food in the supermarkets.

This can be applied as well in our example earlier for war rivalry between countries turning into sports & football rivalry. An analogy of that would be that enthusiasm and passion evolved to be channeled to more peaceful venues. To some extent, we can say that about anger, aggression, and violence as well in this case. Yet harmful traits such as violence should eventually be eliminated if proven regularly harmful and follow natural selection mechanisms on emotions.

This raises an interesting question, do emotions evolve through a mechanism comparable to natural selection or even entirely through natural selection, or is it the other way around? In other words, are emotions one of the key factors driving natural selection?

Charles Darwin himself did sort of explored this, as mentioned earlier. In 1872, he published a book, "The Expression of the Emotions in Man and Animals"; for Darwin, emotion had an evolutionary history that could be traced across cultures and species. He analyzed emotional responses in animals. He compared such responses in a variety of species. He proposed that expressions of emotion evolved in an organism to signal to other organisms what it is about to do next. For example, when an animal feels threatened, it may display a stance of aggression. It may bear its teeth; it may lean forward; it may make aggressive sounds. All of this behavior is to indicate violent behavior, and it does well in intimidating some percentage of other organisms nearby, which may then retreat in fear without necessitating actual combat. Thus this type of behavior has a quantifiable impact on the organism's survival and therefore becomes a preferable trait within the gene pool.

This suggests that emotions could be playing a useful role in natural selection. We could look at it from an epigenetics perspective and understand now that emotions could lead to specific epigenetic patterns, which could eventually lead to changes in the gene pool and mutations.

The approach proposed in this book for quantizing emotions through epigenetics would help in understanding the effect of emotions on natural selection better. Yet as epigenetics itself is still pretty much an emerging and relatively recent field of study, we still have a lot of work to do to get there.

The Universal Collective Impact Behaviors (UCIB)

The need for collective action is one of the EEA properties or the environment of evolutionary adaptiveness, and it has played an important role in our evolution. Collective action and teamwork could be noticed in various groups of animals; good examples are the honey bees and ants building colonies and storing food, even with their limited cognitive abilities. One key I would say evolutionary advantage for humans is that their collective behavior is sort of universal, and it is not limited to a small group of individuals or limited locations or for limited tasks.

This evolutionary advantage for humans in this specific behavior is actually one of the most important traits that helped in our survival and development. It deserves to be emphasized and studied in more detail. Let's give it a name, the UCIB or the Universal Collective Impact Behaviors.

This book will propose three main behaviors of the UCIB or (the Universal Collective Impact Behaviors), which are mainly knowledge-oriented and driven behaviors. Further study and research could reveal more UCIB behaviors.

The knowledge-oriented and driven behaviors:

1.   Seeking knowledge.
2.   Sharing Knowledge.
3.   Implementing knowledge. (Invention & innovation)

These three UCIB behaviors are an essential part of our evolutionary roots, and they are somewhat sort of evolving. The question is, how can we measure and enhance them? Let's first take a look at what do they represent, what could be their properties and a bit of how they evolved.

Some of the properties of the UCIB or (the Universal Collective Impact Behaviors) could be:

1. The unique universality.
2. Curiosity-driven.
3. Ambitions generating.

One of the things that make humans different from other animals is language. We have the cognitive ability to communicate not just about what is happening now but also about what did happen and what could happen. We can think about the future, plan for the future, and imagine the future. Imagination is another key trait that gave us an evolutionary advantage.

There were no chimpanzees around seven million years ago, and there were no humans. But there were CHLCAs (Chimpanzee-Human Last Common Ancestor). Like us, CHLCAs did not have significant natural offenses or defenses like incredible speeds, protective shells, claws, fangs, or venoms. So the CHLCAs were living in the trees' safety till some of them decided to venture down to the savanna. Without appropriate physical abilities, things like cooperation, imagining new strategies, and assigning roles were necessary for survival, all of which is easier if you have a rich collection of symbols that can refer to things across time that developed to be language. The cognitive trade-off hypothesis proposed by Testuro Matsuzawa suggests that we have made a tradeoff between language at the expense of memory ability. In a sense, chimpanzees do not need language and planning for the future, as they basically live in a world of here and now to deal with the daily and immediate situations and threats. This could be mainly the kind of knowledge chimpanzees and other monkeys, for example, need. Seeking more knowledge, sharing knowledge, and implementing knowledge on a high level is not as urgent or required as detailed short-term memory.

Think about all the things that need to happen for a human settlement to thrive: obtaining food, building shelters, raising children, and more. There needs to be a way to divide resources, organize major tasks and distribute labor efficiently. This requires a high level of sophisticated planning and imagination for the future. As societies develop, planning strategies and tools develop. Seeking knowledge, sharing knowledge, and implementing knowledge (especially in the form of innovation and invention) became a crucial part of our evolution. These behaviors could have evolved from a more basic environment of evolutionary adaptiveness (EEA) properties like the emergence of caring and compassion, flattening of social hierarchy, and the need for collective action.

We have seen complex social organization in other species, though, ants, for example, building colonies containing different types of members who perform specific roles. However, this organization does not arise from any higher-level decisions or planning, but it is part of a biologically programmed cycle.

A key property of the UCIB or (the Universal Collective Impact Behaviors) -seeking knowledge, sharing knowledge, and implementing knowledge, especially in the form of innovation and invention- which differentiates humans from other species with a complex social organization is the unique universality of these behaviors. They are not limited to a situation or a location. An inspiring quote by Marie Curie is: "After all science is essentially international, and it is only through lack of the historical sense that national qualities have been attributed to it." In various science fields, we still build on basic concepts and theories introduced centuries ago from different locations and civilizations, which is a key advantageous behavior for humans that is not found in any other species.

Another interesting property of the UCIB behaviors is that they are curiosity-driven in some cases, which means they have not become a basic evolutionary need as they used to be; instead, they developed to be curiosity-driven. We have seen scientific institutes and scientists dedicating their whole lives studying and theorizing for ideas and concepts that do not seem to play any role in our basic need for survival. Yet, in most cases, they turn to be highly beneficial to our development. Usually, fundamental discoveries in science are highly celebrated even in some cases; we do not have the complete picture of what they might lead to in the future. After decades or even centuries, they turn out to open unexpected doors for new theories, discoveries, and inventions. The UCIB behaviors seem to slowly shift from their basic biological evolutionary context to a more curiosity-driven context. While curiosity itself could be understood in an evolutionary context, one might argue that even this property could be explained in the shadows of the evolutionary approach to emotions and behavior.

The third property of the UCIB or (the Universal Collective Impact Behaviors) is that they are ambitions-generating. Something which differentiates us as well from other species. Ambitions are unique to humans. And the UCIB seems to ignite and generate ambitions, which again could be explained in the shadows of the evolutionary approach to emotions and behavior or could be considered another unique property beyond the basic biological evolutionary requirements.

Now, coming back to how we can measure this, we need to measure it, quantize it and control it in order to enhance it and fast track it if possible? This question brings us to the next concept introduced in this book, *The Universal Collective Impact Behaviors Index (UCIBI)*.

The Universal Collective Impact Behaviors Index (UCIBI)

The Universal Collective Impact Behaviors (UCIB) are mainly knowledge and learning-oriented. The proposed UCIB are seeking, sharing, and implementing knowledge in the form of invention and innovation have neural effects on us. The neuroscience of learning is a promising topic explored by researchers and experts. To learn anything, you need to create and strengthen neural pathways in your brain. These are function pathways, neurons working together to achieve a goal. The UCIB have hormonal effects on us as well in terms of the hormones released in response to the emotions –which are mostly pretty positive- developed during engaging in a UCIB (Universal Collective Impact Behaviors)

Our target is to track and measure these neurotransmitters and hormones. We will follow the methods proposed in this book. Measure the unique factor and HN-value of participants involved in UCIB.

The Universal Collective Impact Behaviors Index (UCIBI) would be the unique factor -introduced earlier in the book- for UCIB (Universal Collective Impact Behaviors). This means we will measure the hormonal-neural response with pre-defined conditions of getting involved in UCIB. Then multiply it by each individual's unique factor to get the Universal Collective Impact Behaviors Index (UCIBI).

The law of similarities

First, before we start this chapter, I would like to apologize if the title of the chapter might not seem to reflect what is mentioned in it. The law of similarity in psychology is the gestalt grouping law that states that similar elements tend to be perceived as a unified group. Similarity can refer to any number of features, including color, orientation, size, or even motion.

An interesting thought, which is now more than a hundred years old, was as we started probing the structure of the atom, we looked at it and said there is a nucleus and there are electrons orbiting around, which looks like a solar system. We now know that the laws of physics for a solar system are different from the laws of physics that describe what goes on in an atom, despite the similarities between the atom and the solar system. Like they both have a nucleus, in the case of the solar system, it is a star –the sun in our solar system- and in an atom, the center is the nucleus made up of protons and neutrons. A star is also made up of protons and neutrons, mainly hydrogen and helium. Both are composite objects meaning they have a nucleus and objects that orbit around them. Another similarity is that both contain most of their mass in the center. Another similarity is that they have both orbiting objects. In the case of the solar systems, it is called planets or, more generally, just satellites, and in the case of the atom, they are called electrons. They both have the notion of the conservation of angular momentum, which arises from certain types of symmetries like a symmetry around the rotation axis.

We now understand that the macroscopic world responds to classical physics, and the microscopic world responds to quantum physics. Quantum physics actually applies to the whole universe, but on a large enough scale, it looks like classical physics.

The thought of the similarity between the atom and the solar system was sort of based on similarities of shape and motion. Later on, we have had a better understanding of what are the similarities, what are the fundamental differences, and the laws of physics that describe both.

This is comparable to evolution in a sense. Where thoroughly observing the similarities in shape, motion, and body organs between groups of species led Charles Darwin and the early scientists studying evolution to come up with one of the greatest theories in science. Yet, at that time, they did not have the technologies to discover that there are some astonishing similarities between the DNA of species that are likely to have a common ancestor. Later on, as science advances, we will be able to tell the similarities and differences down to the DNA level.

Another interesting example is actually from String theory, which is referred to as T-duality. The equivalence between theories that look different is usually called a duality. In String theory, we have something called T-duality. It is a duality that relates compactifications on very small geometries to compactifications on big geometries, which surprised everybody when it was first discovered. Geometry can often be recovered or partially recovered from the spectrum of modes or spectrum of vibrations. From the spectrum of particles, the spectrum of vibrational energies, and other kinds of energies, we can predict to a large extent the shape and the size of the compact directions, but there are some dualities that you cannot tell if it is a tiny geometry or if it is a big geometry.

Back to quantum cognition, some psychologists think that a set of ideas borrowed from quantum mechanics could help us make sense of human behavior. The notion is called quantum cognition, suggesting that quantum mechanics' mathematical tools could help make human behavior more predictable. Quantum cognition is building models similar to those of quantum mechanics; these models have proven more successful in predicting human behavior.

The notion here is that these similarities could be more fundamental than just some borrowed ideas and concepts. This is generally the approach in this book and the proposed thought experiments for potentially more fundamental similarities that could help us in our attempt to quantize emotions.

We can use here the analogy of evolution as well. In our case, we are noticing similarities in the effect of emotions on our bodies and some quantum phenomena. We cannot fully understand it, and we are trying to develop the math to better understand these similarities. The notion is that these similarities and building models on a more fundamental level based on them could lead to a much deeper understanding of emotions and quantizing them.

Illusion is half the sickness – Ibn Sina

Ibn Sina (980-1037), a famous scholar and physician who wrote more than 200 works in various disciplines and has been described as the "Father of modern medicine," has two famous quotes referred to him :

- *Illusion is half the sickness, calmness is half the cure, and patience is the first step of healing.*
- *The human mind draws its strength from the strength of the soul, and this is a force that is not insignificant.*

Ibn Sina composed Kitab al-Shifa (Book of the cure), a vast philosophical and scientific encyclopedia, and Al-Qanun fi al-Tibb (The Canon of Medicine), which is among the most famous books in the history of medicine. He is considered one of the first Muslim scholars to talk about the importance of psychotherapy and the impact of mental illness on the nerves and the body, such as fear, anxiety, and sadness.

The first quote for Ibn Sina mentioned here, *"Illusion is half the sickness, calmness is half the cure, and patience is the first step of healing."* It is quite relevant to what we are discussing in this book. We have mentioned earlier the effects of stress on the body and a brief about the long list of chronic stress diseases.

Modern research shows that increased stress, whether from illusion or other reasons, can actually lead to a longer recovery time. One famous experiment demonstrating this involved eleven dental students who volunteered to have their mouths biopsied twice. First during summer vacation and then again during exam week. It took an average of three days longer for the wounds to heal while they were stressed about exams. All kinds of other studies have gotten similar results, some by punching small holes in people as they did with the dental students and others by observing how stress affects recovery from surgery or other major wounds.

What I am proposing in this book is the potential methods to measure and quantize how stress or illusion and calmness could affect the recovery through the HN-value and epigenomes approaches.

Illusion and stress lead to negative hormonal-neural response and epigenetic changes, and specific epigenome patterns. Those changes could lead to disease or a long recovery time.

And vice versa, calmness and patience lead to a positive hormonal-neural response, specific epigenetic changes, and specific epigenome patterns. That helps in faster recovery and even avoiding disease.

The second quote for Ibn Sina mentioned above *"The human mind draws its strength from the strength of the soul and this is a force that is not insignificant."* We have discussed the two-system view of the stress response earlier in this book and how it is a basis for the proposed HN-value (Hormonal-neural response value). We have discussed the hypothalamic-pituitary-adrenocortical axis (HPA axis) and its stress response.

Yet, Ibn Sina`s quote does sort of suggesting another layer that potentially acts as another medium or a translator for emotions to the brain, which he referred to as "the soul." Whether that what he had meant or not, the quote does inspire this line of thought in a sense that the soul in Ibn Sina`s quote could act as the "Field of emotions" proposed earlier in this book, and then the analogy continues as explained in the section of "energy of emotions." He also used the word force in his quote, which is quite interesting again; he might not have actually meant it, but it is sort of relevant for the purpose of our analogy.

Chapter 05

Mathematical psychology & psychophysics

Mathematical psychology & Psychophysics

Mathematical psychology is an approach to psychological research that is based on mathematical modeling of perceptual, thought, cognitive, and motor processes and on the establishment of law-like rules that relate quantifiable stimulus characteristics with quantifiable behavior. The mathematical approach is used to derive hypotheses that are more exact and yield stricter empirical validations. Quantifiable behavior is, in practice, often constituted by task performance.

The application of math in psychology could be traced back to at least the seventeenth century when scientists like Kepler and Galileo were investigating mental process laws. At that time, psychology had not even been recognized as an independent science subject. The applications of mathematics in psychology can be roughly classified into two areas: the mathematical modeling of psychological theories and experimental phenomena, which leads to mathematical psychology. The other is the statistical approach of quantitative measurement practices in psychology, leading to psychometrics.

There are five major research areas in mathematical psychology:

1. learning and memory
2. Perception and psychophysics.
3. Choice and decision-making.
4. Language and thinking.
5. Measurement and scaling.

As quantification of behavior is fundamental in this endeavor, the theory of measurement is a central topic in mathematical psychology. Mathematical psychology is therefore closely related to psychometrics. However, where psychometrics is concerned with individual differences (or population structure) in mostly static variables, mathematical psychology focuses on process models of perceptual, cognitive, and motor processes inferred from the 'average individual.'

Furthermore, where psychometrics investigates the stochastic dependence structure between variables as observed in the population, mathematical psychology almost exclusively focuses on modeling data obtained from experimental paradigms and is therefore even more closely related to experimental psychology/cognitive psychology/psychonomics. Like computational neuroscience and econometrics, mathematical psychology theory often uses statistical optimality as a guiding principle, assuming that the human brain has evolved to solve problems in an optimized way. Central themes from cognitive psychology; limited vs. unlimited processing capacity, serial vs. parallel processing, etc., and their implications are central in rigorous analysis in mathematical psychology.

Mathematical psychologists are active in many fields of psychology, especially in psychophysics, sensation, and perception, problem-solving, decision-making, learning, memory, and language, collectively known as cognitive psychology, and the quantitative analysis of behavior but also, e.g., in clinical psychology, social psychology, and psychology of music.

Mathematics and Psychology before the 19th Century

Choice and decision-making theory is rooted in the development of probability theory. Blaise Pascal considered situations in gambling and further extended to Pascal's wager. In the 18th century, Nicolas Bernoulli proposed the St. Petersburg Paradox in decision making, Daniel Bernoulli gave a solution, and Laplace proposed a modification to the solution later on. In 1763, Bayes published the paper "An Essay towards solving a Problem in the Doctrine of Chances," which is the Bayesian statistics milestone.

Mathematics and Psychology in the 19th Century

The research developments in Germany and England in the 19th century made psychology a new academic subject. Since the German approach emphasized experiments in investigating the psychological processes that all humans share and the England approach measured individual differences, mathematics applications are also different.

In Germany, Wilhelm Wundt established the first experimental psychology laboratory. The math in German psychology is mainly applied in sensory and psychophysics. Ernst Weber (1795–1878) created the first mathematical law of the mind, Weber's law, based on various experiments. Gustav Fechner (1801–1887) contributed math theories in sensations and perceptions, and one of them is Fechner's law, which modifies Weber's law.

Mathematical modeling has a long history in psychology, starting in the 19th century with Ernst Weber (1795–1878) and Gustav Fechner (1801–1887) being among the first to apply the successful mathematical technique of functional equations from physics to psychological processes. They thereby established the fields of experimental psychology in general and psychophysics.

Researchers in astronomy in the 19th century were mapping distances between stars by denoting the exact time of a star's passing of a cross-hair on a telescope. These time measurements relied entirely on human response speed for the lack of the modern era's automatic registration instruments. It had been noted that different astronomers measured small systematic differences in the times, and these were first systematically studied by German astronomer Friedrich Bessel (1782–1846). Bessel constructed *personal equations* from basic response speed measurements that would cancel individual differences from the astronomical calculations. Independently, physicist Hermann von Helmholtz measured reaction times to determine nerve conduction speed, developed the resonance theory of hearing and the Young-Helmholtz theory of color vision.

These two lines of work came together in the research of Dutch physiologist F. C. Donders and his student J. J. de Jaager, who recognized the potential of reaction times for more or less objectively quantifying the amount of time elementary mental operations required. Donders envisioned the employment of his *mental chronometry* to scientifically infer the elements of complex cognitive activity by measurement of *simple reaction time*

Although there are developments in sensation and perception, Johann Herbart developed a mathematical theory system in the cognitive area to understand consciousness's mental process.

The origin of English psychology can be traced to Darwin's theory of evolution. But the emergence of English psychology is because of Francis Galton, who is interested in individual differences between humans on psychological variables. The math in English psychology is mainly statistics, and the work and methods of Galton are the foundation of psychometrics.

Galton introduced bivariate normal distribution in modeling the same individual traits; he also investigated measurement error and built his own model. He also developed a stochastic branching process to examine family names' extinction. A tradition of interest in studying intelligence in English psychology started from Galton. James McKeen Cattell and Alfred Binet developed tests of intelligence.

The first psychological laboratory was established in Germany by Wilhelm Wundt, who amply used Donders' ideas. However, findings that came from the laboratory were hard to replicate, and this was soon attributed to the method of introspection that Wundt introduced. Some of the problems resulted from individual differences in response speed found by astronomers. Although Wundt did not seem to take an interest in these individual variations and focused on studying the *general human mind*, Wundt's U.S. student James McKeen Cattell was fascinated by these differences and started to work on them during his stay in England.

The failure of Wundt's method of introspection led to the rise of different schools of thought. Wundt's laboratory was directed towards conscious human experience, in line with Fechner and Weber's work on stimuli' intensity. In the United Kingdom, under the influence of the anthropometric developments led by Francis Galton, interest focused on individual differences between humans on psychological variables, in line with the work of Bessel. Cattell soon adopted the methods of Galton and helped to lay the foundation of psychometrics.

20th century

Many statistical methods were developed even before the 20th century: Charles Spearman invented factor analysis, which studies individual differences by the variance and covariance. German psychology and English psychology have been combined and taken over by the United States. The statistical methods dominated the field during the beginning of the century. There are two important statistical developments: Structural Equation Modeling (SEM) and analysis of variance (ANOVA). Since factor analysis is unable to make causal inferences, the method of structural equation modeling was developed by Sewall Wright to correlational data to infer causality, which is still a major research area today. Those statistical methods formed psychometrics. The Psychometric Society was established in 1935, and the journal Psychometrika was published in 1936.

In the United States, behaviorism arose in opposition to introspectionism and associated reaction-time research and turned the focus of psychological research entirely to learning theory. In Europe, introspection survived in Gestalt psychology. Behaviorism dominated American psychology until the end of the Second World War and largely refrained from inference on mental processes. Formal theories were mostly absent (except for vision and hearing).

During the war, developments in engineering, mathematical logic and computability theory, computer science and mathematics, and the military need to understand human performance and limitations, brought together experimental psychologists, mathematicians, engineers, physicists, and economists. Out of this mix of different disciplines, mathematical psychology arose. Especially the developments in signal processing, information theory, linear systems, and filter theory, game theory, stochastic processes, and mathematical logic gained a large influence on psychological thinking.

Two seminal papers on learning theory in *Psychological Review* helped establish the field in a world where behaviorists still dominated: A paper by Bush and Mosteller instigated the linear operator approach to learning and a paper by Estes that started the stimulus sampling tradition in psychological theorizing. These two papers presented the first detailed formal accounts of data from learning experiments.

Mathematical modeling of the learning process was greatly developed in the 1950s as the behavioral learning theory was flourishing. One development is the stimulus sampling theory by Williams K. Estes; the other is linear operator models by Robert R. Bush and Frederick Mosteller.

Signal processing and detection theory are broadly used in perception, psychophysics, and nonsensory area of cognition. Von Neumann's book The Theory of Games and Economic Behavior establish the importance of game theory and decision making. R. Duncan Luce and Howard Raiffa contributed to the choice and decision-making area.

The area of language and thinking comes into the spotlight with the development of computer science and linguistics, especially information theory and computation theory. Chomsky proposed the model of linguistics and computational hierarchy theory. Allen Newell and Herbert Simon proposed the model of humans solving problems. Artificial intelligence and human-computer interface development are active in computer science and psychology.

Before the 1950s, psychometricians emphasized the structure of measurement error and the development of high-power statistical methods for the measurement of psychological quantities, but little of the psychometric work concerned the structure of the psychological quantities being measured or the cognitive factors behind the response data. Scott and Suppes studied the relationship between data structure and the structure of numerical systems that represent the data.

Coombs constructed the respondent's formal cognitive models in a measurement situation rather than statistical data processing algorithms, for example, the unfolding model. Another breakthrough is the development of a new form of the psychophysical scaling function along with new methods of collecting psychophysical data, like Stevens' power law.

The 1950s saw a surge in psychological processes' mathematical theories, including Luce's theory of choice, Tanner and Swets' introduction of signal detection theory for human stimulus detection, and Miller's approach to information processing. By the end of the 1950s, the number of mathematical psychologists had increased from a handful by more than tenfold, not counting psychometricians. Most of these were concentrated at Indiana University, Michigan, Pennsylvania, and Stanford. Some of these were regularly invited by the U.S. Social Science Research Council to teach in summer workshops in mathematics for social scientists at Stanford University, promoting collaboration.

To better define the field of mathematical psychology, the mathematical models of the 1950s were brought together in a sequence of volumes edited by Luce, Bush, and Galanter: Two readings and three handbooks. This series of volumes turned out to be helpful in the development of the field. In the summer of 1963, the need was felt for a journal for theoretical and mathematical studies in all areas in psychology, excluding work that was mainly factor analytical. An initiative led by R. C. Atkinson, R. R. Bush, W. K. Estes, R. D. Luce, and P. Suppes resulted in the appearance of the first issue of the Journal of Mathematical Psychology in January 1964.

Under the influence of computer science, logic, and language theory developments, in the 1960s, modeling gravitated towards computational mechanisms and devices. Examples of the latter constitute so-called cognitive architectures (e.g., production rule systems, ACT-R) and connectionist systems or neural networks.

Important mathematical expressions for relations between physical characteristics of stimuli and subjective perception are Weber–Fechner law, Ekman's law, Stevens's power law, Thurstone's law of comparative judgment, the theory of signal detection (borrowed from radar engineering), the matching law, and Rescorla–Wagner rule for classical conditioning. While the first three laws are all deterministic, later established relations are more fundamentally stochastic. This has been a general theme in the evolution in mathematical modeling of psychological processes: from deterministic relations as found in classical physics to inherently stochastic models.

Important theories and models

Sensation, perception, and psychophysics

- Stevens' power law
- Weber–Fechner law

Stimulus detection and discrimination

- Signal detection theory

Stimulus identification

- Accumulator models

- Diffusion models
- Neural network/connectionist models
- Race models
- Random walk models
- Renewal models

Simple decision

- Cascade model
- Level and change race model
- Recruitment model
- SPRT
- Decision field theory

Memory scanning, visual search

- Push-down stack
- Serial exhaustive search (SES) model

Error response times

- Fast guess model

Sequential effect

- Linear operator model

Learning

- Linear operator model
- Stochastic learning theory

Measurement theory

- Theory of conjoint measurement

Psychophysics

Psychophysics quantitatively investigates the relationship between physical stimuli and the sensations and perceptions they produce. Psychophysics has been described as "the scientific study of the relation between stimulus and sensation" or, more completely, as "the analysis of perceptual processes by studying the effect on a subject's experience or behavior of systematically varying the properties of a stimulus along one or more physical dimensions."

Psychophysics also refers to a general class of methods that can be applied to study a perceptual system. Modern applications rely heavily on threshold measurement, ideal observer analysis, and signal detection theory.

Psychophysics has widespread and important practical applications. For example, in the study of digital signal processing, psychophysics has informed the development of models and methods of lossy compression. These models explain why humans perceive very little loss of signal quality when audio and video signals are formatted using lossy compression.

History

Many of the classical techniques and theories of psychophysics were formulated in 1860 when Gustav Theodor Fechner in Leipzig published *Elemente der Psychophysik (Elements of Psychophysics)*. He coined the term "psychophysics," describing research intended to relate physical stimuli to consciousness contents such as sensations *(Empfindungen)*. As a physicist and philosopher, Fechner aimed at developing a method that relates matter to the mind, connecting the publicly observable world and a person's privately experienced impression of it. His ideas were inspired by experimental results on the sense of touch and light obtained in the early 1830s by the German physiologist Ernst Heinrich Weber in Leipzig, most notably those on the minimum discernible difference in intensity of stimuli of moderate strength (just noticeable difference; jnd) which Weber had shown to be a constant fraction of the reference intensity, and which Fechner referred to as Weber's law. From this, Fechner derived his well-known logarithmic scale, now known as the Fechner scale. Weber's and Fechner's work formed one of the psychology bases as a science, with Wilhelm Wundt founding the first laboratory for psychological research in Leipzig (Institut für experimentelle Psychologie). Fechner's work systematized the introspectionist approach (psychology as the science of consciousness), which had to contend with the Behaviorist approach in which even verbal responses are as physical as the stimuli.

During the 1930s, when psychological research in Nazi Germany essentially came to a halt, both approaches eventually began to be replaced by the use of stimulus-response relationships as evidence for conscious or unconscious processing in the mind. Fechner's work was studied and extended by Charles S. Peirce, who was aided by his student Joseph Jastrow, who soon became a distinguished experimental psychologist in his own right. Peirce and Jastrow largely confirmed Fechner's empirical findings, but not all. In particular, Peirce and Jastrow's classic experiment rejected Fechner's estimation of a threshold of perception of weights as being far too high. In their experiment, Peirce, and Jastrow, in fact, invented randomized experiments: They randomly assigned volunteers to a blinded, repeated-measures design to evaluate their ability to discriminate weights. Peirce's experiment inspired other researchers in psychology and education, which developed a research tradition of randomized experiments in laboratories and specialized textbooks in the 1900s. The Peirce–Jastrow experiments were conducted as part of Peirce's application of his pragmaticism program to human perception; other studies considered the perception of light, etc. Jastrow wrote the following summary: "Mr. Peirce's courses in logic gave me my first real experience of intellectual muscle. Though I promptly took to the laboratory of psychology when that was established by Stanley Hall, it was Peirce who gave me my first training in the handling of a psychological problem, and at the same time stimulated my self-esteem by entrusting me, then fairly innocent of any laboratory habits, with a real bit of research. He borrowed the apparatus for me, which I took to my room, installed at my window, and with which, when conditions of illumination were right, I took the observations. The results were published over our joint names in the *National Academy of Sciences Proceedings*. The demonstration that traces of sensory effect too slight to make any registry in consciousness could nonetheless influence judgment may itself have been a persistent motive that induced me years later to undertake a book on *The Subconscious*." This work clearly distinguishes observable cognitive performance from the expression of consciousness.

Modern approaches to sensory perception, such as research on vision, hearing, or touch, measure the perceiver's judgment extracts from the stimulus, often putting aside the question of what sensations are being experienced. One leading method is based on signal detection theory, developed for cases of very weak stimuli. However, the subjectivist approach persists among those in the tradition of Stanley Smith Stevens (1906–1973). Stevens revived the idea of a power-law suggested by 19th-century researchers, in contrast with Fechner's log-linear function (cf. Stevens' power law). He also advocated the assignment of numbers in ratio to stimuli' strengths, called magnitude estimation. Stevens added techniques such as magnitude production and cross-modality matching. He opposed the assignment of stimulus strengths to points on a line labeled in order of strength. Nevertheless, that sort of response has remained popular in applied psychophysics. Such multiple-category layouts are often misnamed Likert scaling after the question items used by Likert to create multi-item psychometric scales, e.g., seven phrases from "strongly agree" through "strongly disagree."

Omar Khaleefa has argued that the medieval scientist Alhazen should be considered the founder of psychophysics. Although Al-Haytham made many subjective reports regarding vision, there is no evidence so far that he used quantitative psychophysical techniques.

Thresholds

Psychophysicists usually employ experimental stimuli that can be objectively measured, such as pure tones varying in intensity or lights varying in luminance. All the senses have been studied: vision, hearing, touch (including skin and enteric perception), taste, smell, and the sense of time. Regardless of the sensory domain, there are three main areas of investigation: absolute thresholds, discrimination thresholds, and scaling.

A threshold (or limen) is the point of intensity at which the participant can just detect the presence of a stimulus (absolute threshold) or the presence of a difference between two stimuli (difference threshold).

Stimuli with intensities below the threshold are considered not detectable (hence: sub-liminal). Stimuli at values close enough to a threshold will often be detectable some proportion of occasions; therefore, a threshold is considered to be the point at which a stimulus, or change in a stimulus, is detected some proportion p of occasions.

Detection

An absolute threshold is the intensity of a stimulus at which the subject can detect the presence of the stimulus some proportion of the time (a p level of 50% is often used). An example of an absolute threshold is the number of hairs on the back of one's hand that must be touched before it can be felt – a participant may be unable to feel a single hair being touched but may be able to feel two or three as this exceeds the threshold. The absolute threshold is also often referred to as the *detection threshold.* Several different methods are used for measuring absolute thresholds (as with discrimination thresholds).

Discrimination

A difference threshold (or just-noticeable difference, JND) is the magnitude of the smallest difference between two stimuli of differing intensities that the participant is able to detect some proportion of the time (the percentage depending on the kind of task). Several different methods are used to test this threshold. The subject may be asked to adjust one stimulus until it is perceived as the same as the other (method of adjustment), may be asked to describe the direction and magnitude of the difference between two stimuli, or may be asked to decide whether intensities in a pair of stimuli are the same or not (forced-choice). The just-noticeable difference (JND) is not a fixed quantity; rather, it depends on how intense the stimuli being measured are and the particular sense being measured. Weber's Law states that the just-noticeable difference of a stimulus is a constant proportion despite variation in intensity.

In discrimination experiments, the experimenter seeks to determine at what point the difference between two stimuli, such as two weights or two sounds, is detectable. The subject is presented with one stimulus, for example, a weight, and is asked to say whether another weight is heavier or lighter (in some experiments, the subject may also say the two weights are the same). At the point of subjective equality (PSE), the subject perceives the two weights to be the same.

The just-noticeable difference, or difference limen (DL), is the magnitude of the difference in stimuli that the subject notices some proportion p of the time (50% is usually used for p in the comparison task). In addition, a two-alternative forced-choice (2-AFC) paradigm can be used to assess the point at which performance reduces to chance on discrimination between two alternatives (p will then typically be 75% since p=50% corresponds to chance in the 2-AFC task).

Absolute and difference thresholds are sometimes considered similar in principle because there is always background noise interfering with our ability to detect stimuli.

Experimentation

In psychophysics, experiments seek to determine whether the subject can detect a stimulus, identify it, differentiate between it and another stimulus, or describe the magnitude or nature of this difference.

Classical psychophysical methods

Psychophysical experiments have traditionally used three methods for testing subjects' perception in stimulus detection and difference detection experiments: the method of limits, the method of constant stimuli, and the adjustment method.

Method of limits

In the ascending method of limits, some property of the stimulus starts out at a level so low that the stimulus could not be detected, then this level is gradually increased until the participant reports that they are aware of it. For example, if the experiment is testing the minimum amplitude of sound that can be detected, the sound begins too quietly to be perceived and is made gradually louder. In the descending method of limits, this is reversed. In each case, the threshold is considered to be the level of the stimulus property at which the stimuli are just detected.

In experiments, the ascending and descending methods are used alternately, and the thresholds are averaged. A possible disadvantage of these methods is that the subject may become accustomed to reporting that they perceive a stimulus and may continue reporting the same way even beyond the threshold (the error of habituation). Conversely, the subject may also anticipate that the stimulus is about to become detectable or undetectable and may make a premature judgment (the error of anticipation).

To avoid these potential pitfalls, Georg von Békésy introduced the staircase procedure in 1960 in his study of auditory perception. In this method, the sound starts out audible and gets quieter after each of the subject's responses until the subject does not report hearing it. At that point, the sound is made louder at each step until the subject reports hearing it, at which point it is made quieter in steps again. This way, the experimenter is able to "zero in" on the threshold.

Method of constant stimuli

Instead of being presented in ascending or descending order, in the method of constant stimuli, the levels of a certain property of the stimulus are not related from one trial to the next but presented randomly. This prevents the subject from being able to predict the level of the next stimulus, and therefore reduces errors of habituation and expectation. For 'absolute thresholds' again, the subject reports whether they are able to detect the stimulus. There has to be a constant comparison stimulus with each of the varied levels for' difference thresholds.' Friedrich Hegelmaier described the method of constant stimuli in an 1852 paper. This method allows for a full sampling of the psychometric function but can result in a lot of trials when several conditions are interleaved.

Method of adjustment

In the method of adjustment, the subject is asked to control the stimulus's level and alter it until it is just barely detectable against the background noise or is the same as the level of another stimulus. The adjustment is repeated many times. This is also called the *method of average error*. In this method, the observers themselves control the variable stimulus's magnitude, beginning with a distinctly greater or lesser level than a standard one, and vary it until they are satisfied by the two's subjective equality. The difference between the variable stimuli and the standard one is recorded after each adjustment, and the error is tabulated for a considerable series. In the end, the mean is calculated, giving the average error, which can be taken as a measure of sensitivity.

Adaptive psychophysical methods

The classic methods of experimentation are often argued to be inefficient. This is because, in advance of testing, the psychometric threshold is usually unknown, and most of the data are collected at points on the psychometric function that provide little information about the parameter of interest, usually the threshold. Adaptive staircase procedures (or the classical method of adjustment) can be used such that the points sampled are clustered around the psychometric threshold. Data points can also be spread in a slightly wider range if the psychometric function's slope is also of interest. Adaptive methods can thus be optimized for estimating the threshold only, or both threshold *and* slope. Adaptive methods are classified into staircase procedures and Bayesian, or maximum-likelihood, methods. Staircase methods rely on the previous response only and are easier to implement. Bayesian methods take the whole set of previous stimulus-response pairs into account and are generally more robust against lapses in attention. Practical examples are found here.

Staircases usually begin with a high-intensity stimulus, which is easy to detect. The intensity is then reduced until the observer makes a mistake, at which point the staircase 'reverses,' and intensity is increased until the observer responds correctly, triggering another reversal. The values for the last of these 'reversals' are then averaged. There are many different types of staircase procedures using different decision and termination rules. Step-size, up/down rules, and the spread of the underlying psychometric function dictate where they converge on the psychometric function. Threshold values obtained from staircases can fluctuate wildly, so care must be taken in their design. Many different staircase algorithms have been modeled and some practical recommendations suggested by Garcia-Perez.

One of the more common staircase designs (with fixed-step sizes) is the 1-up-N-down staircase. If the participant makes the correct response N times in a row, the stimulus intensity is reduced by one step size. If the participant makes an incorrect response, the stimulus intensity is increased by one size. A threshold is estimated from the mean midpoint of all runs. This estimate approaches, asymptotically, the correct threshold.

Bayesian and maximum-likelihood procedures

Bayesian and maximum-likelihood (ML) adaptive procedures behave, from the observer's perspective, similar to the staircase procedures. The choice of the next intensity level works differently; however: After each observer response, from the set of this and all previous stimulus/response pairs, the likelihood is calculated of where the threshold lies. The point of maximum likelihood is then chosen as the best estimate for the threshold, and the next stimulus is presented at that level (since a decision at that level will add the most information). A prior likelihood is further included in the calculation in a Bayesian procedure. Compared to staircase procedures, Bayesian and ML procedures are more time-consuming to implement but are considered to be more robust. This kind's well-known procedures are Quest, ML-PEST, and Kontsevich & Tyler's method.

Magnitude estimation

In the prototypical case, people are asked to assign numbers in proportion to the magnitude of the stimulus. This psychometric function of the geometric means of their numbers is often a power law with a stable, replicable exponent. Although contexts can change the law & exponent, that change too is stable and replicable. Instead of numbers, other sensory or cognitive dimensions can be used to match a stimulus, and the method then becomes "magnitude production" or "cross-modality matching." The exponents of those dimensions found in numerical magnitude estimation predict the exponents found in magnitude production. Magnitude estimation generally finds lower exponents for the psychophysical function than multiple-category responses because of the restricted range of the categorical anchors, such as those used by Likert as items in attitude scales.

Since this is not a textbook, I will not be presenting here these theories and models of mathematical psychology and psychophysics in further detail, yet in the indices, I will be adding some references for the main theories and pioneers of mathematical psychology & psychophysics.

The model introduced in this book is building on previous research from various related fields of science, including mathematical psychology & psychophysics. Additionally, it applies quantum cognition methodologies to build a new model to quantize emotions.

Let`s review the main concepts introduced in this book:

1. The in sign ⵣ introduced earlier in this book.
2. Considering the wave function as a complicated potential as well leading to introducing this equation:
   $i\hbar\frac{\partial}{\partial t}\psi(x,t) = \hat{E} \divideontimes \hat{\psi}(x,t)$
3. The freedom to swap rules between the energy operator and the wave function operatee.
4. The five main principles for the basic mathematics of psychology :
   a. The unique factor.
   b. Pre-defined conditions.
   c. Imposed conditions.
   d. Complex states.
   e. Isolation principle.
5. The equation of energy in shape giving a certain function
   $\hat{E} \divideontimes \hat{s} = f\hat{\Psi}(x,t)$
6. The equation of energy in pattern giving a certain function
   $\hat{E} \divideontimes \hat{P} = f\hat{\Psi}(x,t)$
7. The freedom to swap the "IN product," the operator, and the operatee. In other words, the energy operator could act on the wave function to give a certain shape or topological manifold ....etc in that case, the equations would be : $\hat{E} \divideontimes f\hat{\Psi}(x,t) = \hat{s}$ & $\hat{E} \divideontimes f\hat{\Psi}(x,t) = \hat{P}$
8. The HN-value (Hormonal-Neural value) as a method to quantize emotions.

9. The process of quantizing emotions, in an attempt to visualize it :

Color, shape, sound ⚹ ➤ Hormones / Neurotransmitters
Measure & quantize

10. An approach to quantize emotions through epigenetics and checking which emotions or rather patterns of emotions (P) will lead to which epigenome pattern.
   A simple form would be: EP ⚹ DNA = EPP

11. The Universal Collective Impact Behaviors (UCIB).

12. The Universal Collective Impact Behaviors Index (UCIBI).

There are other new concepts introduced in this book that are more analogous:

1. The Standard Model & periodic table of emotions analogy.
2. "The particle of emotion – The Ematon" analogy.
3. The fields of emotions and the particle-like emotions.
4. Controlling the evolution of emotions.

Chapter 06

Epigenetics as a good candidate

Epigenomes as a good candidate

Epigenomes are a very good candidate that could assist us in quantizing emotions. Yet, since it is a relatively recent field of study, I preferred to rely on the HN-Value (The hormonal, neural response) in this book.

Let's have a quick brief about epigenomes and epigenetics. This will help us understand better how the science of epigenetics could be a good candidate to quantize emotions.

 A famous example that is usually used to explain epigenetics and how epigenomes work is identical twins with the same DNA—and studying how they turn out after years so different. For example, one of the twins could be having diseases, and the other could be healthy. The science of Epigenetics studies how DNA interacts with the multitude of smaller molecules in the cells, which can activate and deactivate genes. They answer an interesting question: how come each of the approximately 200 cell types in your body has essentially the same genome, yet they perform different functions?

Another analogy that is usually used to explain epigenetics is that you think of DNA as a recipe book. The epigenomes determine what gets cooked when. They are not making any conscious choices themselves; rather, their presence and concertation within the cells make the difference. Genes in DNA are expressed when they are read and transcribed into RNA, which is translated into proteins by structures called ribosomes. And proteins are much of what determines the cell's characteristics and function. Epigenetic changes can boost or interfere with the transcription of specific genes. The most common way interference happens is that DNA or the proteins it's wrapped around get labeled with small chemical tags. The set of all of the chemical tags that are attached to the genome of a given cell is called the epigenome. Like the methyl group, some of these inhibit gene expression by derailing the cellular transcription machinery or causing the DNA to coil more tightly, making it inaccessible. The gene is still there, but it is silent.

Boosting transcription is essentially the opposite. Some chemical tags will unwind the DNA, making it easier to transcribe, which ramps up the associated protein production. Epigenetic changes can survive cell division, which means that they could affect an organism for its entire life.

The epigenomes also mediate a lifelong dialogue between genes and the environment. The epigenomes act as a turn-on, turn-off switch. They are chemical tags that turn genes on and off, and they can be influenced by factors including diet, chemical exposure, smoking, medication, etc. The resulting epigenetic changes can eventually lead to disease if, for example, they turn off a gene that makes a tumor-suppressing protein.

Environmentally-induced epigenetic changes are part of the reason why genetically identical twins can grow up to have very different lives. As twins get older, their epigenomes diverge, affecting the way they age and their disease susceptibility.

Social experiences are another significant reason for epigenetic changes, which will partially explain why it could be an excellent candidate when quantizing emotions.

In one famous experiment when mother rats were not attentive enough to their pups, genes in the babies that helped them manage stress were methylated and turned off.

The epigenome is changeable. One of the biggest influences on your epigenome, as you get older, is your environment, so direct influences such as your diet and indirect environmental changes such as stress. According to these influences, you will have a different epigenetic pattern. You will have different genes that are switched on and different genes which are switched off.

Understanding the patterns of epigenomes is an emerging field of study. I will mention a general overview of how studying epigenomes' patterns could be a great tool for quantizing emotions; then, I will get a little more technical.

Identical twins are a great resource to study the influences of epigenetic changes. And to put this into the context of what I am trying to propose. We need to isolate all the factors that could lead to epigenetic changes and limit them to the emotional or psychological factors (which is not, of course, an easy task to do)

The idea is that we expose our identical twins to identical environmental conditions, identical diets, physical activities, and sports, etc. Of course, there will be a margin of error. Still, we should try to be as accurate as we could get. The only difference in the factors that could lead to epigenetic changes should be the factor we want to study. Let`s say happiness versus stress or even different levels of the same emotion. So lots of happiness and joy versus average or sort of seasonal happy events.

The experiment goes that we get a particular pair of chromosomes from the identical twins over specific periods, let`s say 3 to 6 months or potentially longer spans of time. For example, we get the methylation pattern for one of the twins colored in one color and for the other one colored in a different color. Then we overlay the images to observe the identical and the different methylation or targeted epigenetic changes. We can observe the changes in the patterns through a specific period of time.

The idea is that if we managed to understand which specific emotions and even which specific level of emotions could lead to which specific patterns of epigenomes, we will be having a direct tool for observing and even measuring the effect of emotions in a quantifiable way.

This tool could be quite powerful because the changes in the patterns of the epigenomes cause direct physical changes; they can lead to disease or recovery. Having a tool to measure the effect of emotions on these patterns will enable us to design what could be referred to as patterns of emotions that could lead to patterns of epigenomes potentially leading to avoid disease or even help in recovery from disease. It could open the door for further research with very promising potential.

It is worth mentioning that the experiment of getting a pair of chromosomes and getting a pair of chromosomes, for example, the methylation pattern for one of the twins colored in one color and the other colored in a different color. Then overlaying the images to observe the identical and the different methylation or targeted epigenetic changes is not a new experiment being proposed in this book. It is a known experiment to the scientist working in the field of epigenetics. I added some resources and video references for this experiment in the bibliography and references, as this kind of experiment can be very helpful in utilizing epigenetics in our attempt to quantize emotions.

There is a good example of how nutrition influences the epigenome, which could inspire more ideas on how emotions could affect the epigenome and isolate specific factors leading to epigenetic changes and observe the results.

It is the example found in bees. Queen bees and worker bees are genetically identical, they have exactly the same DNA, and the only difference is that the queen bees are force-fed royal jelly from the minute they are at the larval stage. The worker bees are fed on nectar, pollen, and water. The fact that the queen bees are fed on royal jelly switches on genes in her that helped her develop ovaries and a really large abdomen for laying eggs and gives her a queenly attitude, which makes the other worker bees work under her command. The DNA between these two types of bees is identical, and the only difference between them is that the queen bees get fed royal jelly. So her diet is switching on particular genes to develop her ovaries in her abdomen, and the worker bees remain sterile. That is entirely epigenetic changes related to diet.

Again this is not a textbook, so I will not be explaining in detail about epigenetics. I will be adding in the indices some references and resources. Yet, in the next couple of pages, I will get just a little bit more technical about how do epigenomes work.

The way it works starts largely with the DNA winding and the proteins that are used to do that and compact the DNA into the cells. The first level of DNA winding is the helical shape of the DNA that starts to compact it a bit, but in order to get it right inside the nucleus, the DNA is wound around proteins called histones, and histones are wound around one another.

So the first type of epigenetic modification we have is called histone modifications, and those take place on the histone proteins' tails. Each of the histone tails has various points at which you can add different chemical signals. There are a number of different chemicals you can add to the tails, for example, acetyl groups, phosphate groups, methyl groups, and ubiquinone groups.

The position of each of these tags on the tail and whatever is lying next to it, greatly influences what these particular chemical tags do. For example, DNA methylation always turns off gene expression, but it is not that clear when it comes to histones; you can have methyl groups in particular positions that will unwind the DNA and help gene expression. If you put them in a different position, it will wind up the DNA and turn off gene expression. Generally, if the histone tail is acetylated, DNA is unwound, exposes the genes to the transcription machinery, and increases expression. It gets much more complicated when talking about the methyl groups, the phosphate groups, and the ubiquinone groups. Those particular groups' positions, where they are, and how they interact can either open or close the DNA.

And that is an emerging field of epigenetics trying to find which combinations will open and which combinations will close. The second type of modification happens to the genes themselves, to which I will add some relevant references and resources in the indices.

So, as epigenetics study develops and we have a better understanding of which combinations or patterns will open and which combinations or patterns will close, along with the reasons that lead to these specific patterns being added to these specific histone tails. Potentially we can figure out the underlying mathematics behind it.

A very primitive and basic attempt would be to check which emotions or patterns of emotions (P) will lead to an epigenome pattern.

A simple form would be: $EP \divideontimes DNA = EPP$

Where EP is the emotions pattern "in" the DNA leading to the EPP (Epigenome pattern),

$$\begin{vmatrix} E_1 & E_2 \\ E_3 & E_4 \end{vmatrix} \divideontimes DNA = \begin{vmatrix} AG & MG \\ PG & UG \end{vmatrix}$$

Where E_1, E_2, E_3 & E_4 represent a pattern of four emotions operating on a DNA leading to a pattern of the epigenome, which is AG for acetyl groups, PG for phosphate groups, MG for methyl groups, and UG for ubiquinone groups.

We will still need to know more information about the epigenome pattern's location and the duration in which this pattern is attached to our genes or histone tails. This could be done by getting a pair of chromosomes from the participant and observing the epigenome patterns. Doing this over specific periods of time, we can come to know as well the duration these patterns stayed in these locations.

In a sense, we would need to know the epigenome pattern as a function or description of their location/position and duration/time. I would propose to write this as follows:

$$\begin{vmatrix} E_1 & E_2 \\ E_3 & E_4 \end{vmatrix} \divideontimes DNA = \begin{vmatrix} AG & MG \\ PG & UG \end{vmatrix}_{(x,t)}$$

Where x describes the location/position of the epigenome pattern and t describes the duration/time these patterns stayed in these locations.

When we say DNA here for more sophisticated and targeted results, we probably would need to run genome sequencing or know the gene sequence of the part we are testing. This kind of equation would of course, require computer-aided calculations and potentially computer simulations and is not the kind of equation that could be solved traditionally on a whiteboard or a notebook.

This equation would as well allow us to apply the five main principles introduced earlier:

1. The unique factor.
2. Pre-defined conditions.
3. Imposed conditions.
4. Complex states.
5. Isolation principle.

The unique factor: As explained earlier, the unique factor would be a relativistic factor/constant that differs from one operatee to another. In our epigenome story, the unique factor could affect the epigenomes' patterns. And the way to verify that is by exposing two pairs of identical twins to the exact same conditions and observing the difference in the epigenome patterns between them. If the epigenome patterns are identical, then probably the unique factor does not play an important role in forming the epigenome patterns. But if there are variations, in that case, we would conclude that every person –even identical twins- is indeed having a unique factor that could be verified.

Pre-defined conditions: A group of conditions specific to each operation that should be pre-defined to provide the necessary information to carry out the operation and predefine the required information to be shown in the "IN product" of the operation. In our epigenome story, the pre-defined conditions are quite important. We have to specify which factors we want to pre-define, unite, or differentiate, to observe their influence on the epigenome patterns and gene modifications.

Imposed conditions: A condition or more that is unconventional or unexpected, when imposed on the operation, will lead to a change in the results (the IN product). Here as well, the imposed conditions are important in our epigenome story to observe which imposed conditions would lead to which epigenome patterns and changes.

Complex states: In this one, we have a number of factors/states that we need to consider, and we require to know more information from the IN product. This concept could be applied to our epigenome story when measuring complex situations. We could observe the epigenome changes due to complex factors as nutrition, environment, emotions, etc.

The isolation principle: It is the principle of isolating specific states or conditions from the operation to get the IN product's required pre-defined results. Which defiantly would play a very important role in our epigenome story. In terms of which emotions or patterns of emotions, we want to isolate and even which epigenome patterns we want to isolate. For example, if we want just to measure the (AG) acetyl groups & (PG) phosphate groups and isolate the (MG) methyl groups and (UG) ubiquinone groups. For example, if you want just to consider the methyl group and so on.

Emotions & Stress-related diseases

Your hearts are like the engine in a car. It has valves, and it has pressurized chambers, and just like the engine of the car is responsible for making the rest of the parts of the car run, your heart is responsible for the performance of a lot of the other organ functions in your body by getting blood and nutrients to them.

The car is designed to accelerate pretty quickly every once in a while, and similarly, your heart and blood vessels respond to acute stress by kicking into overdrive. So your heart rate in the force of contraction increases, and those nutrients and oxygen flow around the body faster in order to support your organs and tissues, and this is good if we are responding to an emergency. But just like we do not want to burn up the engine in the car by driving around at the maximum RPMs all the time, having our hearts constantly operating in overdrive can be pretty damaging.

One of the damaging effects of stress on our hearts is increased blood pressure, which can lead to a disease called hypertension; one of the damaging effects of hypertension might be a vascular disease, which refers to blood vessels disease. And it's not only the heart that could be negatively affected; part of our metabolism -or the process of us breaking down food sources to get energy- can be impacted negatively as well.

When you experience acute stress, your body activates a system called the hypothalamic-pituitary-adrenocortical axis (HPA axis). It starts in your brain in the limbic system (the part responsible for a lot of your automatic emotional reactions, among other things). There a region called the hypothalamus releases hormones that start a whole chain of more hormones being released, first by the pituitary gland and then by the adrenal glands, which release a bunch of adrenaline and cortisol into the bloodstream. And those two hormones trigger the "Fight-or-Flight" response. They boost physical activity by increasing the blood sugar and the blood flow to your muscles and bump up your metabolism at the same time.

The physical boost helps you fight the stressors. The same system is activated by chronic stress, but things get more complicated. Researchers have found that people under some kind of chronic stress have perpetually high cortisol levels as if their HPA axis is constantly running. But we do know that while the stress reaction can be helpful at times, having it running all the time is a problem.

People under chronic stress are at higher risks for all kinds of ailments like heart disease, autoimmune disease, and mental disorders like anxiety and depression. Under chronic stress, the stress response is constantly sapping your energy. The resources used to fight-or-flight have to come from somewhere, and one of the places they come from is the immune system.

On the molecular level, the same cortisol that works to get extra glucose to the muscles also stops the body from making as many infection-fighting white blood cells as it usually would. So stress can sort of tank your ability to fight infections.

Stress can also advance the aging process. By the time people get older, the DNA has had to replicate so many times that the protective parts of each of the ends of the chromosome, called telomeres, can kind of start to fray. When telomeres are shorter, it is more likely that there will be errors in copying genes, and those errors increase the risk of disease. There is evidence that having more cortisol in the blood interprets the repair of the telomeres, which might explain why stress is linked to diseases that are also associated with age, like heart disease, cancer, and anemia. To stay healthy, the best thing you can do is to get rid of the chronic stress, but easier said than done; if you cannot get rid of it completely things like meditation and relaxation therapies can help lower your stress response.

Too much of cortisol results in loss of synaptic connections between neurons and the shrinking of the prefrontal cortex the part of the brain that regulates behaviors like concentration, decision-making, judgement and social interaction. It also leads to fewer new brain cells being made in the hippocampus.

It is not all bad news, though; there are many ways to reverse what cortisol does to a stressed brain. Some of the most powerful weapons are exercise and meditation, which involves breathing deeply and being aware and focused on your surroundings, which decrease stress and increase the size of the hippocampus, thereby improving your memory. So we have to train ourselves to control our stress and prevent it from taking control of us.

Happiness and laughter do positively affect our bodies, including reducing stress and epigenome changes. Our goal in this book is to propose some attempts to measure and quantize the positive and negative effects of emotions on our bodies and try to control them.

The main two proposed approaches in this book could work as well to sort of measure & quantize the effects of chronic stress, for example, in the body, and start controlling it. The two approaches are the HN-Value (Hormonal, neural response) and the using epigenetics. And as mentioned earlier, the HN-value would be sort of a bit more currently mature and developed.

HN-values for stressors and de-stressors.

We could identify the stressors for the patients, categorize them and assign an HN-value for stressors, which will enable us to measure the hormonal-neural response value for every stressor on the patient. Additionally, we could assign HN-value for what could be referred to as the "de-stressors," which would be for the purpose of this book "emotional de-stressors."

Once we are able to determine the HN-values for the stressors and de-stressors, we will be able to control the process and introduce de-stressors with specific values to manage the stress and reduce it to the minimum.

Stimulating positive epigenome patterns.

As the field of epigenetics develops, stimulating positive epigenome patterns could be done as well, as we will have a better understanding of which patterns of epigenomes have positive effects and which patterns have negative effects.

Then we start introducing de-stressors to enhance and stimulate the "positive causing epigenome patterns" and work on avoiding the "negative causing epigenome patterns."

Chapter 07

New concepts introduced & future research

The main new concepts introduced in this book:

Suppose we followed the recent quantum cognition approach in building models drawn from quantum mechanics. In that case, we could consider that the model proposed in this book is technically borrowing the idea of Schrodinger's equation in an attempt to deal with cognitive phenomena and quantize emotions.

In Schrodinger's equation, we do not assume that the energy is a number because we do not know it. In general, if the particle is moving in a complicated potential, you do not know what the possible energies are.

$$i\hbar\frac{\partial}{\partial t}\psi(x,t) = \hat{E}\psi(x,t)$$

This is symbolically what must be happening because if this particle has a definite energy, then the energy operator $\hat{E}$ gives you the energy acting on the function.

So I borrowed this idea to deal with cognitive phenomena and quantize emotions, then I introduced these main concepts to help to build my model, which are basically:

1. The IN sign ✳ introduced earlier in this book.
2. Considering the wave function as a complicated potential as well leading to introducing this equation:
   $$i\hbar\frac{\partial}{\partial t}\psi(x,t) = \hat{E} ✳ \hat{\psi}(x,t)$$
3. The freedom to swap rules between the energy operator and the wave function operatee.
4. The five main principles for the basic mathematics of psychology :
   a. The unique factor.
   b. Pre-defined conditions.
   c. Imposed conditions.
   d. Complex states.

　　　　e. Isolation principle.

5. The equation of energy in shape giving a certain function
$$\hat{E} \ast \hat{s} = f^{\Psi}(x,t)$$

6. The equation of energy in pattern giving a certain function
$$\hat{E} \ast P = f^{\Psi}(x,t)$$

7. The freedom to swap the "IN product," the operator, and the operatee. In other words, the energy operator could act on the wave function to give a certain shape or topological manifold ....etc in that case, the equations would be : $\hat{E} \ast f^{\Psi}(x,t) = \hat{s}$ & $\hat{E} \ast f^{\Psi}(x,t) = P$

8. The HN-value (Hormonal-Neural value) as a method to quantize emotions.

9. The process of quantizing emotions, in an attempt to visualize it:

Color, shape, sound $\quad \ast \blacktriangleright$ Hormones / Neurotransmitters
Measure & quantize

10. An approach to quantize emotions through epigenetics and checking which emotions or rather patterns of emotions (P) will lead to which epigenome pattern.
A simple form would be: EP $\ast$ DNA = EPP

11. The Universal Collective Impact Behaviors (UCIB).

12. The Universal Collective Impact Behaviors Index (UCIBI).

There are other new concepts introduced in this book that are more analogous:

1. The Standard Model & periodic table of emotions analogy.
2. "The particle of emotion – The Ematon" analogy.
3. The fields of emotions and the particle-like emotions.
4. Controlling the evolution of emotions.

The model introduced in this book is proposing that the energy operator is operating on a complicated potential or, in other words, a field of waves, which is fundamentally how energy operators operate; they do not operate on a single wave. They operate on a field of waves.

Basically, considering emotions as an energy operator of complicated potential acting on humans who are considered a field of wave functions of complicated potential with multiple quantum states.

The energy operator requires the freedom to act on the operatee's multiple quantum states. Both the operator and the operatee can affect each other, so this requires the freedom to swap roles as introduced in our thought experiment, "quantizing emotions."

Applications and future research:

How can Psychiatrists use this tool & how to deal with the five main principles?

What is proposed in this book could be very powerful for psychiatrists, a reminder of the five main principles:

1. The unique factor.
2. Pre-defined conditions.
3. Imposed conditions.
4. Complex states.
5. Isolation principle.

This tool could be used with people facing mental illnesses and disorders as well as people who want to improve their mental health. Comparable to your medical record, some tests and experiments could be done to determine the client's HN (Hormonal and Neural responses) & HN-value to various conditions, including happy and sad circumstances. Some experiments could be done to determine the client's unique factor, which should be updated regularly, probably every 3-5 years. But depending on the case, it could be updated annually or even quarterly.

Next, some of these principles can be controlled and adjusted like the unique factor, and some we will have to deal with in different ways. Psychiatrists can train their clients to improve their unique factors by training them to perceive things in different ways and then measure the progress by measuring the changes in the HN (Hormonal and neural responses) and the HN-value. This is not limited to training the clients to perceive unfortunate events with a different positive perspective, but it could be training the clients to see the positive things even in their daily routine that could improve their unique factors. Or even help them discover new habits and practices that could improve their unique factor and validate that by measuring the hormonal and neural responses.

Using the unique factor to measure the effects of the treatments numerically could help a lot in many ways, including clearly monitoring the progress of the client's mental conditions and health.

The pre-defined conditions are defined by us to tackle specific circumstances and measure their effects on the hormonal levels and neural response. This could change depending on the circumstances and issues we want to tackle. It can be used to understand would positively affect the mental health of the client.

The imposed conditions that cause initial disturbance to the operation are obviously out of control of both the clients and the psychiatrists, yet psychiatrists can impose new conditions to the operation to mitigate or enhance the effects of the initial imposed conditions.

Complex states would require deeper understanding and study by psychiatrists. It is not always obvious or easy to fully cover the small and complex events that can serve as catalysts acting on starting conditions and causing changes in the operation. Yet, a good psychiatrist would be able to sort of decode them, and the more details the psychiatrist gets hold of, the better.

The isolation principle could just be a tool to isolate certain circumstances and carry out the operation accordingly. But it could be a powerful tool to isolate some events and circumstances intentionally. A good psychiatrist would be able to understand and isolate the client from the effects of the events that can cause mental disorders.

Applications and future research:

Measuring and controlling the evolution of emotions

Even though this book is mainly addressing how to quantize emotions and develop the basic mathematics for psychology yet, there could be many applications for the concepts proposed in this book that go well beyond psychology. Before we get there, one of the future research goals for this book is the detailed quantizing of emotions through physical equations, potentially Schrodinger's equation. Integrating the "IN sign" in Schrodinger's equation will potentially become a tool to measure and control how emotions evolve in time, comparable to Schrodinger's equation currently describing how the wave function evolves in time.

The escaping graviton

The approach for testing string theory is mainly through particle collisions. Particle collisions allow all sorts of sub-atomic and "extraordinary" particles to be released. The string theory proposes that these collisions could generate the graviton giving evidence of gravity seeping of our membrane universe into "higher dimensions." These collisions should show that the amount of energy after the collisions are less than the amount of energy before the collisions as gravitons have seeped into higher dimensions. But because gravitons are so weak, a detector with almost the mass of Jupiter and 100% efficiency, placed in close orbit around a neutron star, would only be expected to observe one graviton every ten years, even under the most favorable conditions, which is pretty challenging.

We are dealing with the graviton as a particle. And we are trying to measure the decrease in energy/mass after the collision. The "IN sign" could be used to deal with the graviton as a wave; in that case, measuring the change in the "frequency" of the graviton instantly once the collision is done and potentially femtoseconds after the collision will give us evidence this graviton is basically losing all its classical behavior as a particle in our universe and is getting "tuned" with higher dimensions.

So, could this tool be used to quantize gravity?

It is something that could be explored. Yet, if we want to examine if the graviton is escaping to higher dimensions, we might as well try to examine the effects of these higher dimensions on the frequency/wavelength of the graviton. The tool introduced in this book allows the operator and operatee to affect each other and swap roles. Let's consider the graviton as a wave function operated on by energies in higher dimensions. We could potentially detect and measure the effect of the energies from higher dimensions on the graviton's frequency/wavelength.

Energy in shapes & patterns an equation for many applications

The equations introduced in this book where energy in shape gives a certain function are not limited to quantizing emotions. It could be applied to other concepts in physics and mathematics in general.

$$\hat{E} \ast \hat{s} = f^{\Psi}(x,t)$$

Where $\hat{E}$ is the energy operator, $\hat{s}$ is the shape operatee and $f^{\Psi}(x,t)$ is the resulting wave function. This proposed equation prescribes that energy in shape gives a wave function, which could be applied in principle to almost everything.

The concept goes further to the idea of the pattern, where the movement or rather the fluctuations of energy or complicated potential of energy in a particular pattern creates a specific function or a wave function; it could as well written as follows:

$$\hat{E} \ast \hat{P} = f^{\Psi}(x,t)$$

Where $\hat{P}$ refers to the pattern in which the energy operator moves; following the same logic, the operator can affect the operatee and vice versa.

Bibliography and References

Can we consider psychology & quantizing emotions a science?

1. Fanelli, D. (2010). *Positive Results increase down the hierarchy of the sciences* PLOS ONE DOI: 10.1371/journal.pone.0010068.
2. Fredrickson BL, Losada MF. *Positive affect and the complex dynamics of human flourishing*. Am Psychol. 2005 Oct; 60(7):678-86. doi: 10.1037/0003-066X.60.7.678. Erratum in: Am Psychol. 2013 Dec;68(9):822. PMID: 16221001; PMCID: PMC3126111.
3. Scishow Psych, *Is Psychology a Science?* Scishow Psych YouTube Channel, March; 2017.
4. Taylor SE, & Lobel M (1989). *Social comparison activity under threat: downward evaluation and upward contacts*. Psychological Review, 96 (4), 569-75 PMID: 2678204.

Chapter one: Thought experiments.

5. Alfred North Whitehead and Bertrand Russel, *Principia Mathematica*, First and second editions, Cambridge at the University Press, 1910-1913 & 1925-1927.
6. Bernhardt PC, Dabbs JM Jr, Fielden JA, Lutter CD. *Testosterone changes during vicarious experiences of winning and losing among fans at sporting events*. Physiol Behav. 1998 Aug; 65(1):59-62. doi: 10.1016/s0031-9384(98)00147-4. PMID: 9811365.
7. Oliveira, Tania & Gouveia, Maria & Oliveira, Rui. (2009). *Testosterone responsiveness to winning and losing experiences in female soccer players*. Psychoneuroendocrinology. 34. 1056-64. 10.1016/j.psyneuen.2009.02.006.

Chapter Two: The proposed mathematics

8. Schoppe Oliver, Harper Nicol S., Willmore Ben D. B., King Andrew J., Schnupp Jan W. H, *Measuring the Performance of Neural Models, Frontiers in Computational Neuroscience*, 2016.

9.   AUTHOR=Gu Simeng, Wang Fushun, Patel Nitesh P., Bourgeois James A., Huang Jason H, *A Model for Basic Emotions Using Observations of Behavior in Drosophila*, Frontiers in Psychology, 2019.

10. June Gruber, *Human Emotion 2.3: Emotion measurement*, YaleCourses YouTube Channel, Yale University, May; 2013.

11. Panawala, Lakna. (2017). *Difference between Hormones and Neurotransmitters*, Research gate.

12. Crashcourse, *The Chemical Mind: Crash Course Psychology #3*, Crashcourse YouTube Channel, Feb; 2014.

13. Brambilla DJ et al. *Intraindividual variation in levels of serum testosterone and other reproductive and adrenal hormones in men*. Clin Endocrinol (Oxf). 2007;67:853-862.

14. Mishra RG et al. *Metabolite ligands of estrogen receptor-beta reduce primate hyperreactivity*. Am J Physiol. 2006;290(1):H295-303.

15. Granger DA. et al. *The "trouble" with salivary testosterone. Psychoneuroendocrinology*. 2004 Nov;29(10):1229-40.

16. Pushpa Larsen ND, Michael Kaplan ND, Leah Alvarado ND, and Mi-Jung Lee ND, LAC, *Hormone testing: When to use serum, saliva and urine*, Meridian Valley Lab.

17. Ryan Donovan, Aoife Johnson, Aine de Roiste, Ruairi O`Reilly, Cork Institute of technology, *Quantifying the links between personality, sub-traits and the basic emotions*, Research YouTube Channel, August; 2020.

18. Greg J. Norman, Elizabeth Necka, Gary G. Berntson, *The Psychophysiology of Emotions*, Science direct, Editor(s): Herbert L. Meiselman, Emotion Measurement, Woodhead Publishing, 2016.

19. Wallace, M. (2011-07). *Measurement of hormones*. In Oxford Textbook of Endocrinology and Diabetes. Oxford, UK: Oxford University Press. Retrieved 12 Feb. 2021.

20. Professor Dave Explains, *The Psychology of Emotions and Stress*, Professor Dave Explains YouTube Channel, January; 2020.

21. Roesler WJ, Park EA. *Hormone response units: one plus one equals more than two*. Mol Cell Biochem. 1998 Jan;178(1-2):1-8. doi: 10.1023/a:1006886421795. PMID: 9546576.

22. Pucci E, Franchi F, Kicovic PM, Sgrilli R, Barletta D, Argenio GF, Gasperi M, Bernini GP, *Luisi M. Amplification of LH response to LHRH by dopamine infusion in eugonadal women*. J Endocrinol Invest. 1981 Jan-Mar;4(1):55-8. doi: 10.1007/BF03349415. PMID: 6787110

23. Chernecky CC, Berger BJ. Catecholamines - plasma. In: Chernecky CC, Berger BJ, eds. *Laboratory Tests and Diagnostic Procedures*. 6th ed. St Louis, MO: Elsevier Saunders; 2013:302-305.

24. Guber HA, Farag AF, Lo J, Sharp J. *Evaluation of endocrine function. In: McPherson RA, Pincus MR*, eds. *Henry's Clinical Diagnosis and Management by Laboratory Methods.* 23rd ed. St Louis, MO: Elsevier; 2017:chap 24.

25. Young WF. *Adrenal medulla, catecholamines, and pheochromocytoma.* In: Goldman L, Schafer AI, eds. *Goldman-Cecil Medicine.* 25th ed. Philadelphia, PA: Elsevier Saunders; 2016:chap 228.

Chapter Three: The proposed quantum mechanical approach & applications

26. Barton Zwiebach, *Operators and the Schrödinger Equation*, MIT 8.04 Quantum Physics I, spring 2013, MIT OpenCourseWare, License: Creative Commons BY-NC-SA.

27. Leonard Susskind, *Lecture 9 | String Theory and M-Theory*, Stanford YouTube Channel, Stanford, November; 2010.

28. Professor Dave Explains, *Higher Derivatives and Their Applications*, Professor Dave Explains YouTube Channel, March; 2018.

29. wikipedia.org/wiki/Emotion_classification#*Plutchik's_wheel_of_emotions*, License: Creative Commons BY-SA-3.0.GFDL.

30. Plutchik, Robert (1980), Emotion: *Theory, research, and experience*: Vol. 1. Theories of emotion, **1**, New York: Academic

31. Plutchik, Robert (2002), *Emotions and Life: Perspectives from Psychology, Biology, and Evolution*, Washington, DC: American Psychological Association

32. Plutchik, Robert; *R. Conte., Hope* (1997), *Circumplex Models of Personality and Emotions*, Washington, DC: American Psychological Association

33. LondonCityGirl, *String Theory Made Simple*, LondonCityGirl YouTube Channel, January; 2018.

34. Ekman, Paul (1992). *An Argument for Basic Emotions. Cognition and Emotion*. 6(3/4): 169–200.

35. Ramamurti Shankar, *24. Quantum Mechanics VI: Time-dependent Schrödinger Equation*, Fundamentals of Physics, II (PHYS 201), Spring, 2010, YaleCourses YouTube Channel.

36. Harry Cliff, *Beyond the Higgs: What's Next for the LHC? - with Harry Cliff*, October, 2017, The Royal Institution YouTube Channel.

37. wikipedia.org/wiki/*Crowd_psychology*. License: Creative Commons BY-SA-3.0.GFDL.

38. Dimitri Nanopoulos, *What is a particle?,* Quanta Magazine, Editor: Natalie Wolchover, November; 2020.

39. Leonard Susskind, *Lecture 6 | Quantum Entanglements, Part 1 (Stanford)*, Stanford YouTube Channel, Stanford, Fall; 2006.
40. Helen Quinn, *What is a particle?,* Quanta Magazine, Editor: Natalie Wolchover, November; 2020.
41. Mary Gaillard, *What is a particle?,* Quanta Magazine, Editor: Natalie Wolchover, November; 2020.
42. David Tong, *Quantum Fields: The Real Building Blocks of the Universe - with David Tong*, February 2017, The Royal Institution YouTube Channel.
43. Roger Penrose, *The Emperor's New Mind: Concerning Computers, Minds and The Laws of Physics*, 1989.
44. Scishow Psych, *Studying the Brain with Quantum Mechanics*, Scishow Psych YouTube Channel, September; 2020.
45. V.I. Yukalov and D. Sornette, *Decision theory with prospect interference and entanglement*, Springer Science + Business Media, Feb; 2010, doc rero digital library.
46. Harald Atmanspacher, Thomas Filk, and Hartmann R¨omer, *Quantum Zeno Features of Bistable Perception*, Cornell University, Feb; 2003.
47. Professor Dave Explains, *Quantum Numbers, Atomic Orbitals, and Electron Configurations*, Professor Dave Explains YouTube Channel, August; 2015.
48. Nicoletta Lanese, *What is quantum cognition? Physics theory could predict human behavior*, Live Science, 2020.
49. Zheng Wanga, Tyler Sollowaya, Richard M. Shiffrinb, and Jerome R. Busemeyerb, *Context effects produced by question orders reveal quantum nature of human judgments*, School of Communication, The Ohio State University, Columbus, OH 43210; and b Department of Psychological and Brain Sciences, Indiana University, Bloomington, IN 47405. PNAS, July; 2014.
50. Jim Al-Khalili, *How Quantum Biology Might Explain Life's Biggest Questions | Jim Al-Khalili | TED Talks*, TED YouTube Channel, September; 2015.

Chapter Four: Evolution of emotions

51. June Gruber, *Human Emotion 4.1: Evolution and Emotion I (Introduction)*, YaleCourses YouTube Channel, Yale University, May; 2013.
52. Charles Darwin, *The Expression of the Emotions in Man and Animals*, Publisher: John Murray, 1872.
53. Paul Ekman, *Darwin, and Facial Expression: A Century of Research in Review*, July; 2015.

54. Paul Ekman and Richard J. Davidson, *The Nature of Emotion: Fundamental Questions*, 1994.
55. PBS Eons, *How Dogs (Eventually) Became Our Best Friends*, PBS Eons YouTube Channel, Yale University, April; 2020.
56. Jane J. Lee, *Dog and Human Genomes Evolved Together*, National Geographic website, May; 2013.
57. Robert Wayne, Anna Kukekova, and Robert Franciscus, CARTA: *Domestication: Transformation of Wolf to Dog; Fox Domestication; Craniofacial Feminization*, University of California Television (UCTV) YouTube Channel, November; 2014.
58. PBS Eons, *How We Domesticated Cats (Twice)*, PBS Eons YouTube Channel, Yale University, November; 2019.
59. Ian Morris, *War! What Is It Good For?: Conflict and the Progress of Civilization from Primates to Robots,* April; 2014.
60. Vsauce, *The Cognitive Tradeoff Hypothesis,* Vsauce YouTube Channel, December; 2018.
61. Bruce McCandliss, *The Neuroscience of Learning - Bruce McCandliss*, Stanford YouTube Channel, Stanford, November; 2015.
62. Jeanette Norden, *The Neuroscience of Learning and Memory,* Vanderbilt University YouTube Channel, March; 2014.
63. Barkhoudarian G, Hovda DA, Giza CC. *The molecular pathophysiology of concussive brain injury*. Clin Sports Med. 2011; 30(1):
64. Giza CC, Hovda DA. *The new neurometabolic cascade of concussion. Neurosurgery*. 2014; 75 Suppl 4:S24-33.
65. Neuroscientifically Challenged, *2-Minute Neuroscience: Concussions,* Neuroscientifically Challenged YouTube Channel, February; 2019.

Chapter Five: Mathematical psychology & psychophysics

66. Estes, W. K. (2001-01-01), *"Mathematical Psychology, History of"*, in Smelser, Neil J.; Baltes, Paul B. (eds.), International Encyclopedia of the Social & Behavioral Sciences, Pergamon, pp. 9412–9416, *doi:10.1016/b0-08-043076-7/00647-1, ISBN 978-0-08-043076-8, retrieved 2019-11-23*
67. McKenzie, James (2020), *"Pascal's wager"*, Wikipedia, 33 (3), p. 21, *Bibcode:2020PhyW...33c..21M, doi:10.1088/2058-7058/33/3/24, retrieved 2019-11-24*
68. Leahey, T. H. (1987). **A History of Psychology (Second ed.).** Englewood Cliffs, NJ: Prentice Hall. *ISBN 0-13-391764-9.*
69. Batchelder, W. H. (2002). *"Mathematical Psychology"*. In Kazdin, A. E. (ed.). *Encyclopedia of Psychology*. Washington/NY: APA/Oxford University Press. *ISBN 1-55798-654-1.*

70. Mosteller, F. (1951). "*A mathematical model for simple learning*." Psychological Review. 58 (5): 313–323. *doi:10.1037/h0054388. PMID 14883244.*

71. Estes, W. K. (1950). "*Toward a statistical theory of learning*." Psychological Review. 57 (2): 94–107. *doi:10.1037/h0058559.*

72. Scott, Dana; Suppes, Patrick (June 1958). "*Foundational aspects of theories of measurement1*". The Journal of Symbolic Logic. 23 (2): 113–128. *doi:10.2307/2964389. ISSN 0022-4812. JSTOR 2964389.*

73. Coombs, Clyde H. (1950). "*Psychological scaling without a unit of measurement*." Psychological Review. 57 (3): 145–158. *doi:10.1037/h0060984. ISSN 1939-1471. PMID 15417683.*

74. Stevens, S. S. (1957). "**On the psychophysical law.**" Psychological Review. 64 (3): 153–181. *doi:10.1037/h0046162. ISSN 1939-1471. PMID 13441853.*

75. *Luce, R. D., Bush, R. R. & Galanter, E. (Eds.) (1963). **Readings in mathematical psychology.** Volumes I & II. New York: Wiley.*

76. *Luce, R. D., Bush, R. R. & Galanter, E. (Eds.) (1963). **Handbook of mathematical psychology**. Volumes I-III. New York: Wiley. Volume II from Internet Archive*

77. Luce, R. Duncan (1986). **Response Times: Their Role in Inferring Elementary Mental Organization**. Oxford Psychology Series. 8. New York: Oxford University Press. *ISBN 0-19-503642-5.*

78. https://en.wikipedia.org/wiki/**Mathematical_psychology**, License: Creative Commons BY-SA-3.0.GFDL.

79. Gescheider G (1997). **Psychophysics: the fundamentals. Somatosensory & Motor Research**. **14** (3rd ed.). pp. 181–8. doi:10.1080/08990229771042. ISBN 978-0-8058-2281-6. PMID 9402648.

80. Bruce V, Green PR, Georgeson MA (1996). **Visual perception (3rd ed.).** Psychology Press.

81. Boff KR; Kaufman L; Thomas JP (eds.). **Handbook of perception and human performance: Vol. I. Sensory processes and perception**. New York: John Wiley. PMID 9402648.

82. Gescheider G (1997). "**Chapter 5: The Theory of Signal Detection". Psychophysics: the fundamentals (3rd ed.).** Lawrence Erlbaum Associates. ISBN 978-0-8058-2281-6. PMID 9402648.

83. Gustav Theodor Fechner (1860). **Elemente der Psychophysik (Elements of Psychophysics).**

84. Snodgrass JG. 1975. **Psychophysics. In: Experimental Sensory Psychology**. B Scharf. (Ed.) pp. 17–67.

85. Gescheider G (1997). *"Chapter 1: Psychophysical Measurement of Thresholds: Differential Sensitivity"*. *Psychophysics: the fundamentals (3rd ed.)*. Lawrence Erlbaum Associates. ISBN 978-0-8058-2281-6. PMID 9402648.

86. Broadbent DE. 1964. Behavior; Neisser U. 1970. *Cognitive psychology.*

87. Charles Sanders Peirce and Joseph Jastrow (1885). *"On Small Differences in Sensation"*. Memoirs of the National Academy of Sciences. **3**: 73–83.

88. Hacking, Ian (September 1988). *"Telepathy: Origins of Randomization in Experimental Design."* Isis. **79** (3, "A Special Issue on Artifact and Experiment"): 427–451. doi:10.1086/354775. JSTOR 234674. MR 1013489. S2CID 52201011.

89. Stephen M. Stigler (November 1992). *"A Historical View of Statistical Concepts in Psychology and Educational Research."* American Journal of Education. **101** (1): 60–70. doi:10.1086/444032. S2CID 143685203.

90. Trudy Dehue (December 1997). "Deception, Efficiency, and Random Groups*: Psychology and the Gradual Origination of the Random Group Design*" (PDF). Isis. **88**(4): 653–673. doi:10.1086/383850. PMID 9519574. S2CID 23526321.

91. Omar Khaleefa (1999). *"Who Is the Founder of Psychophysics and Experimental Psychology?"*. American Journal of Islamic Social Sciences. **16** (2).

92. Aaen-Stockdale, C.R. (2008). *"Ibn al-Haytham and psychophysics."* Perception. **37** (4): 636–638. doi:10.1068/p5940. PMID 18546671. S2CID 43532965.

93. Gescheider G (1997). "Chapter 2*: Psychophysical Measurement of Thresholds: Absolute Sensitivity"*. Psychophysics: the fundamentals (3rd ed.). Lawrence Erlbaum Associates. ISBN 978-0-8058-2281-6. PMID 9402648.

94. John Krantz. *"Experiencing Sensation and Perception"* Archived 2017-11-17 at the Wayback Machine. pp. 2.3–2.4. Retrieved May 29, 2012.

95. Gustav Theodor Fechner (1860). *Elemente der Psychophysik (Elements of Psychophysics),* Kap. IX: Das Weber'sche Gesetz.

96. *Psychology: the Science of Behaviour*. 4th ED. Neil R. Carlson, C. Donald Heth

97. Gescheider G (1997). *"Chapter 4: Classical Psychophysical Theory"*. *Psychophysics: the fundamentals (3rd ed.)*. Lawrence Erlbaum Associates. ISBN 978-0-8058-2281-6. PMID 9402648.

98. Strasburger H (1995–2020). *Software for visual psychophysics: an overview*. VisionScience.com

99. Gescheider G (1997). *"Chapter 3: The Classical Psychophysical Methods". Psychophysics: the fundamentals (3rd ed.).* Lawrence Erlbaum Associates. ISBN 978-0-8058-2281-6. PMID 9402648.

100. Donald; Janet Laming (1992). *"F. Hegelmaier: On memory for the length of a line."* Psychological Research. **54** (4): 233–239. doi:10.1007/BF01358261. ISSN 0340-0727. PMID 1494608. S2CID 6965887.

101. Treutwein, Bernhard (September 1995). *"Adaptive psychophysical procedures".* Vision Research. **35** (17): 2503–2522. doi:10.1016/0042-6989(95)00016-X. PMID 8594817. S2CID 10550300.

102. Garcia-Perez, MA (1998). *"Forced-choice staircases with fixed step sizes: asymptotic and small-sample properties".* Vision Res. **38** (12): 1861–81. doi:10.1016/S0042-6989(97)00340-4. PMID 9797963. S2CID 18832392.

103. Watson, Andrew B.; Pelli, Denis G. (March 1983). **"Quest: A Bayesian adaptive psychometric method".** Perception & Psychophysics. 33 (2): 113–120. doi:10.3758/BF03202828. PMID 6844102.

104. Harvey, Lewis O. (November 1986). **"Efficient estimation of sensory thresholds".** *Behavior Research Methods, Instruments, & Computers.* 18 (6): 623–632. doi:10.3758/BF03201438.

105. Kontsevich, Leonid L.; Tyler, Christopher W. (August 1999). *"Bayesian adaptive estimation of psychometric slope and threshold."* Vision Research. 39 (16): 2729–2737. doi:10.1016/S0042-6989(98)00285-5. PMID 10492833. S2CID 8464834.

106. Stevens, S. S. (1957). *"On the psychophysical law."* Psychological Review. 64 (3): 153–181. doi:10.1037/h0046162. PMID 13441853.

Chapter Six: Epigenetics as a good candidate

107. TED-Ed, *What is epigenetics? - Carlos Guerrero-Bosagna*, TED-Ed YouTube Channel, June; 2016.

108. Jemma Berry, *Introducing epigenetics*, spiceuwa YouTube Channel, March; 2016.

109. khanacademymedicine, *Physical effects of stress | Processing the Environment | MCAT | Khan Academy*, khanacademymedicine YouTube Channel, April; 2014.

110. Maria Juarez-Reyes, *Beyond Stress, and Anxiety: How Stress Affects the Body and What You Can Do to Manage It*, Stanford Health Care YouTube Channel, August; 2018.

111. American Psychological Association. (2019, October 25). *Stress won't go away? Maybe you are suffering from chronic stress.* http://www.apa.org/topics/stress/chronic

112. Epel ES, Blackburn EH, Lin J, Dhabhar FS, Adler NE, Morrow JD, Cawthon RM. *Accelerated telomere shortening in response to life stress*. Proc Natl Acad Sci U S A. 2004 Dec 7;101(49):17312-5. doi: 10.1073/pnas.0407162101. Epub 2004 Dec 1. PMID: 15574496; PMCID: PMC534658.

113. Vincent Walsh, Joe Herbert, Shane O'Mara and Julie Turner-Cobb, *The Science of Stress: From Psychology to Physiology*, The Royal Institution YouTube Channel, July; 2017.

114. Calado RT, Young NS. *Telomere diseases*. *N Engl J Med*. 2009;361(24):2353-2365. doi:10.1056/NEJMra0903373

115. Scishow Psych, *How Chronic Stress Harms Your Body*, Scishow Psych YouTube Channel, December; 2017.

116. Webster Marketon JI, Glaser R. *Stress hormones and immune function*. Cell Immunol. 2008 Mar-Apr;252(1-2):16-26. doi: 10.1016/j.cellimm.2007.09.006. Epub 2008 Feb 14. PMID: 18279846.

117. Marucha PT, Kiecolt-Glaser JK, Favagehi M. *Mucosal wound healing is impaired by examination stress*. Psychosom Med. 1998 May-Jun;60(3):362-5. doi: 10.1097/00006842-199805000-00025. PMID: 9625226.

118. Cohen S. *Social status and susceptibility to respiratory infections*. Ann N Y Acad Sci. 1999;896:246-53. doi: 10.1111/j.1749-6632.1999.tb08119.x. PMID: 10681901.

119. Miller GE, Chen E, Zhou ES. *If it goes up, must it come down? Chronic stress and the hypothalamic-pituitary-adrenocortical axis in humans.* Psychol Bull. 2007 Jan;133(1):25-45. doi: 10.1037/0033-2909.133.1.25. PMID: 17201569.

120. Jackson JS, Knight KM, Rafferty JA. *Race and unhealthy behaviors: chronic stress, the HPA axis, and physical and mental health disparities over the life course*. *Am J Public Health*. 2010;100(5):933-939. doi:10.2105/AJPH.2008.143446

121. Flannery, Michael. "*Avicenna.*" *Encyclopedia Britannica*, 1 Jan. 2021, https://www.britannica.com/biography/Avicenna. Accessed 12 February 2021.

122. Ibn Sina, *Kitab al-Shifa (Book of the cure)*

123. Ibn Sina, *Al-Qanun fi al-Tibb (The Canon of medicine)*

124. Cano, Miguel Ángel et al. *"Depressive symptoms and externalizing behaviors among Hispanic immigrant adolescents: Examining longitudinal effects of cultural stress."* Journal of adolescence vol. 42 (2015): 31-9. doi:10.1016/j.adolescence.2015.03.017

125. Tomaka, J., Blascovich, J., Kelsey, R. M., & Leitten, C. L. (1993). *Subjective, physiological, and behavioral effects of threat and challenge appraisal.* Journal of Personality and Social Psychology, 65(2), 248–260. https://doi.org/10.1037/0022-3514.65.2.248

126. Harvey A, Nathens AB, Bandiera G, Leblanc VR. *Threat and challenge: cognitive appraisal and stress responses in simulated trauma resuscitations.* Med Educ. 2010 Jun;44(6):587-94. doi: 10.1111/j.1365-2923.2010.03634.x. PMID: 20604855.

127. Gouin JP, Kiecolt-Glaser JK. *The impact of psychological stress on wound healing: methods and mechanisms.* Immunol Allergy Clin North Am. 2011;31(1):81-93. doi:10.1016/j.iac.2010.09.010

128. Stojanovich L, Marisavljevich D. *Stress as a trigger of autoimmune disease.* Autoimmun Rev. 2008 Jan;7(3):209-13. doi: 10.1016/j.autrev.2007.11.007. Epub 2007 Nov 29. PMID: 18190880.

Index

Science is the crawling of humanity towards the truth.

This book does not claim to be proposing any scientific theories. It is proposing two main approaches to quantize emotions. It starts with basic and intuitive thought experiments to build up the logic used to propose these approaches.

Some of the ideas proposed in the book:

- An attempt to quantize emotions through the Hormonal-Neural response based on the two system view of stress response.
- Proposing a value for the Hormonal-Neural response that could be referred to as the HN-value.
- Suggesting that the science of epigenetics could be another approach to quantize emotions.
- Proposing five main principles when attempting to quantize emotions:
  - The unique factor.
  - Pre-defined conditions.
  - Imposed conditions.
  - Complex states.
  - Isolation principle.
- Proposing what could be referred to as the Universal Collective Impact Behaviors (UCIB)

The book used quantum cognition methodologies to build analogies that could help in the attempt to quantize emotions.

Thought experiments and analogies had been the most successful approach of humanity that led to the most significant discoveries in the history of humankind. Quantizing emotions (The basic mathematics of psychology) is unleashing scientific imagination without claiming any theories. It is opening doors for further scientific research and validation.